Springer Theses

Recognizing Outstanding Ph.D. Research

For further volumes:
http://www.springer.com/series/8790

Aims and Scope

The series "Springer Theses" brings together a selection of the very best Ph.D. theses from around the world and across the physical sciences. Nominated and endorsed by two recognized specialists, each published volume has been selected for its scientific excellence and the high impact of its contents for the pertinent field of research. For greater accessibility to non-specialists, the published versions include an extended introduction, as well as a foreword by the student's supervisor explaining the special relevance of the work for the field. As a whole, the series will provide a valuable resource both for newcomers to the research fields described, and for other scientists seeking detailed background information on special questions. Finally, it provides an accredited documentation of the valuable contributions made by today's younger generation of scientists.

Theses are accepted into the series by invited nomination only and must fulfill all of the following criteria

- They must be written in good English.
- The topic should fall within the confines of Chemistry, Physics, Earth Sciences, Engineering and related interdisciplinary fields such as Materials, Nanoscience, Chemical Engineering, Complex Systems and Biophysics.
- The work reported in the thesis must represent a significant scientific advance.
- If the thesis includes previously published material, permission to reproduce this must be gained from the respective copyright holder.
- They must have been examined and passed during the 12 months prior to nomination.
- Each thesis should include a foreword by the supervisor outlining the significance of its content.
- The theses should have a clearly defined structure including an introduction accessible to scientists not expert in that particular field.

Henning Döscher

GaP Heteroepitaxy on Si(100)

Benchmarking Surface Signals when Growing GaP on Si in CVD Ambients

Doctoral Thesis accepted by
Humboldt Universität zu Berlin, Germany

Springer

Author
Dr. Henning Döscher
Institut für Physik
Technische Universität Ilmenau
Ilmenau
Germany

Foreword by
Prof. Kerstin Volz
Struktur- & Technologieforschungslabor
Philipps-Universität Marburg
Marburg
Germany

ISSN 2190-5053 ISSN 2190-5061 (electronic)
ISBN 978-3-319-37955-5 ISBN 978-3-319-02880-4 (eBook)
DOI 10.1007/978-3-319-02880-4
Springer Cham Heidelberg New York Dordrecht London

Printed on acid-free paper

Springer is part of Springer Science+Business Media (www.springer.com)

Related Journal Articles

1. H. Döscher, T. Hannappel, B. Kunert, A. Beyer, K. Volz, and W. Stolz: '*In situ verification of single-domain III–V on Si(100) growth* via *metal–organic vapor phase epitaxy*', Applied Physics Letters **93** [17] (2008), p. 172110
2. H. Döscher, A. Dobrich, S. Brückner, P. Kleinschmidt, and T. Hannappel: '*Si(100) surfaces in a hydrogen-based process ambient*', Applied Physics Letters **97** [15] (2010), p. 151905
3. H. Döscher and T. Hannappel: '*In situ reflection anisotropy spectroscopy analysis of heteroepitaxial GaP films grown on Si (100)*', Journal of Applied Physics **107** [12] (2010), p. 123523
4. H. Döscher, P. Kleinschmidt, and T. Hannappel: '*Atomic surface structure of Si(100) substrates prepared in a chemical vapor environment*', Applied Surface Science **257** [2] (2010), p. 574
5. H. Döscher, B. Kunert, A. Beyer, O. Supplie, K. Volz, W. Stolz, and T. Hannappel: '*In situ anti phase domain quantification applied on hetero-epitaxial GaP growth on Si(100)*', Journal of Vacuum Science & Technology B **28** [4] (2010), p. C5H1
6. H. Döscher, B. Borkenhagen, G. Lilienkamp, W. Daum, and T. Hannappel: '*III–V on silicon: Observation of gallium phosphide anti-phase disorder by low-energy electron microscopyy*', Surface Science Letters **605** [15–16] (2011), p. L38
7. H. Döscher, S. Brückner, A. Dobrich, C. Höhn, P. Kleinschmidt, and T. Hannappel: '*Surface preparation of Si(100) by thermal oxide removal in a chemical vapor environment*', Journal of Crystal Growth **315** [1] (2011), p. 10
8. H. Döscher, O. Supplie, S. Brückner, T. Hannappel, A. Beyer, J. Ohlmann and K. Volz: '*Indirect* in situ *characterization of Si(100) substrates at the initial stage of III–V heteroepitaxy*', Journal of Crystal Growth **315** [1] (2011), p. 16
9. H. Döscher, K. Möller, and T. Hannappel: '*GaP(100) and InP(100) surface structures during preparation in a nitrogen ambient*', Journal of Crystal Growth **318** [1] (2011), p. 372
10. H. Döscher, S. Brückner, and T. Hannappel: '*Investigation of oxide removal from Si(100) substrates in dependence of the MOVPE process gas ambient*', Journal of Crystal Growth **318** [1] (2011), p. 563

11. O. Supplie, M. Pristovsek, T. Hannappel, and H. Döscher,: *'In situ access to the dielectric anisotropy of buried III–V/Si(100) heterointerfaces'*, Physical Review B **86** [3] (2012), p. 035308
12. P. Kleinschmidt, H. Döscher, P. Vogt, and T. Hannappel: *'Direct observation of dimer flipping at the hydrogen-stabilized GaP(100) and InP(100) surfaces'*, Physical Review B **83** [15] (2011), p.155316
13. A. Dobrich, P. Kleinschmidt, H. Döscher, and T. Hannappel: *'Quantitative investigation of hydrogen bonds on Si(100) surfaces prepared by vapor phase epitaxy)'*, Journal of Vacuum Science & Technology B **29** [4] (2011), p. 04D114
14. S. Brückner, H. Döscher, P. Kleinschmidt, O. Supplie, A. Dobrich, and T. Hannappel: *'Anomalous double-layer step formation on Si(100) in hydrogen process ambient'*, Physical Review B **86** [19] (2012), p.195310
15. S. Brückner, H. Döscher, P. Kleinschmidt, and T. Hannappel: *'In situ investigation of hydrogen interacting with Si(100)'*, Applied Physics Letters **98** [21] (2011), p. 211909

Foreword

The integration of III/V semiconductors on silicon substrates has been addressed for many years in semiconductor research. A high quality overgrowth of silicon with III/V semiconductors would allow the design of several novel devices, which are not possible using silicon itself, due to fundamental limitations arising from this semiconductor's band structure. The most promising of the possible applications is highly efficient III/V solar cells on silicon, direct bandgap III/V lasers, and high mobility n-channel layers for III/V on silicon CMOS. Prior to the deposition of an optoelectronically active III/V device layer on silicon, the growth of a buffer layer, into which potential defects can be confined, is advised. Gallium phosphide (GaP) can serve as such a buffer layer, as it has nearly the same lattice constant as silicon and hence one can—at the III/V nucleation stage—deal with all the challenges of merging an isovalent semiconductor with a heterovalent one, except for the lattice mismatch. The relaxation of subsequent III/V layers, if needed, can be done at a later stage of growth. This facilitates the control of defects.

Henning Döscher applied a variety of surface-sensitive techniques to first of all establish a silicon surface, which was prepared in a CVD (chemical vapor deposition) environment, suitable for later III/V growth. Also using methods of surface science, he made important contributions to understanding the MOVPE (metal organic vapor phase epitaxy) growth of GaP better. Very important for the success of his work was the in-situ spectroscopic technique reflectance anisotropy spectroscopy (RAS) , which allowed him to benchmark reflectance signals with signals derived from the same surface using ex-situ surface science tools. This is especially beneficial when gas-phase processes like CVD or MOVPE are used to grow functional layers, as in these techniques there is no direct in-situ access to the surface of the growing film using electron-based analysis techniques.

By carefully applying these methods and also carrying out some theoretical modeling, especially of the RAS signals, Henning Döscher successfully implemented a GaP buffer layer on (001) silicon substrates with different off-orientations. These buffer layers can serve as virtual substrates for subsequent growth of active III/V device structures.

Henning Döscher published his results in numerous peer-reviewed Journal papers and presented them—also in the framework of invited talks—at various international conferences. His contributions to the field were honored by several awards, including the LayTec In Situ Monitoring Award (2008), Young Scientist

Awards by DGKK (2010) and PCSI (2010), the SKM dissertation prize of the DPG (2012), as well as the 'Welt der Zukunft: Energie 2050' for the best essay contributed (2010).

Marburg, August 2013 Prof. Kerstin Volz

Acknowledgments

I want to express my deep gratitude to all those persons, who supported me and my work in various valuable ways, in particular, I would like to mention

- Prof. Dr. Thomas Hannappel for supervision of my work at the Helmholtz Zentrum Berlin, providing both the freedom to develop own ideas and the support to realize them,
- Prof. Dr. Ted Masselink for his evaluation of my thesis at the Humboldt-Universität zu Berlin,
- Prof. Dr. Winfried Daum for his external evaluation as well as for the hospitality in his group at TU Clausthal,
- the whole crew of the Former Department E-I5 for an excellent working atmosphere, many cake sessions, and productive collaboration. Eventually, this department became an extraordinary reservoir of experts in III–V/Si(100) integration,
- Anja Dobrich, Sjoerd Verhagen, and Oliver Supplie for their successful work in their diploma theses exploring Si–H vibrations, InGaP grading, and GaP/Si(100) interface anisotropy, respectively,
- Dr. Peter Kleinschmidt for his mastery regarding both STM tips and English language,
- Sebastian Brückner for tieing in with my thesis, particularly regarding the analysis on XPS and in-situ RAS signals associated with Si(100),
- Christian Höhn and Sven Kubala for enduring care for our MOCVD and UHV systems, for implementation so many modifications, and for their support in the execution of experiments and measurements,
- Ulrike Bloeck for joint effort on dark-field TEM, where Dr. Daniel Abou-Ras and Prof. Kerstin Volz gave us many valuable hints,
- and many others who supported my work as student assistant, as summer student, as intern, as lab guest, or simply as colleagues,
- our collaboration partners for excellent scientific exchange:
- Prof. Dr. Kerstin Volz, Prof. Dr. Wolfgang Stolz, Dr. Bernardette Kunert, Andreas Beyer, Jens Ohlmann, and all others involved in joint quantitative in-situ analysis and benchmarking with AFM and TEM at Phillips Universität Marburg,

- Dr. Gerhard Lilienkamp and Benjamin Borkenhagen in the group of Prof. Daum at TU Clausthal for joint LEEM investigations,
- Dr. Markus Pristovsek (TU Berlin) for enlightening discussions about our RAS calculations,
- Dr. Kristof Möller (Azur Space) and Dr. Patrick Vogt (TU Berlin) for supplying complimentary InP(100) data,
- Dr. Frank Dimroth and Tobias Roesener (ISE) for our joint project work,
- Nils Nüsse, Michael Barth, and Janik Wolters for extraordinary collaboration in the development of photonic crystals based on our GaP/Si(100) layers,
- for helpful discussion, planned collaborations, and many hints: Dr. Holger Eisele, Dr. Tobias Wietler, Dr. Sibylle Gemming, Prof. Dr. Wolf-Gero Schmidt, Dr. Fariba Hatami, Ingo Regolin, and so many others.
- for financial support of my thesis work: BMBF (within the project 'III–V–Si-Solar') as well as DGKK (young academics prize) and PCSI (Young Scientist Award) for supporting conference attendance,
- all my friends, who reminded me over and over again on all the other aspects of life,
- my family and especially my girlfriend, who always supported me by any means available.

Contents

Chapter 1
Introduction

1.1 Motivation: High Efficiency Photovoltaics on Si(100) Substrates

Solar cells enable renewable power supply by direct conversion of solar radiation into electric energy. Light absorption in semiconductor materials generates electron–hole pairs which can be separated by p/n-junctions. In principle, the conversion efficiency is limited by matching the band gap of the utilized semiconductors with the continuous distribution of photon energies in the solar spectrum (Fig. 1.1). Transmission of the radiation occurs below the band gap, while almost immediate thermalization of charge carriers after absorption process essentially confines the yield per photon to the band gap energy. Accordingly, Shockley and Queisser [1] calculated an upper theoretical limit for the conversion efficiency of a conventional single-junction solar cell in the range of 30 %.

As shown in Fig. 1.1, light transmitted through an individual solar cell can still be exploited by another one with a lower band gap, which enables increasing the overall conversion efficiency. Serial connection of multiple single-junction subcells with different band gaps creates so-called multi-junction solar cells which already achieved conversion efficiencies significantly above 40 % under concentrated sunlight [2]. Present high-efficiency multi-junction solar cells typically consist of two different III–V semiconductor sub-cells grown by metal–organic vapor phase epitaxy (MOVPE) on Ge(100) substrates (Fig. 1.2). Striving for large scale terrestrial application of multi-junction solar cells in concentrator photovoltaics, the high abundance of silicon, in particular compared to conventional substrate materials such as gallium arsenide, indium phosphide and germanium is just one source of the desire to utilize Si(100) instead. Due to the superior manufacturing experience associated with the mature microelectronic industry, Si(100) substrates are produced with high purity, large wafer sizes and in significant quantities. In general, silicon features desirable values of crucial material properties such as hardness, mechanical integrity and thermal conductivity, which are less pronounced for classical III–V semiconductors.

H. Döscher, *GaP heteroepitaxy on Si(100)*,
Springer Theses, DOI: 10.1007/978-3-319-02880-4_1,

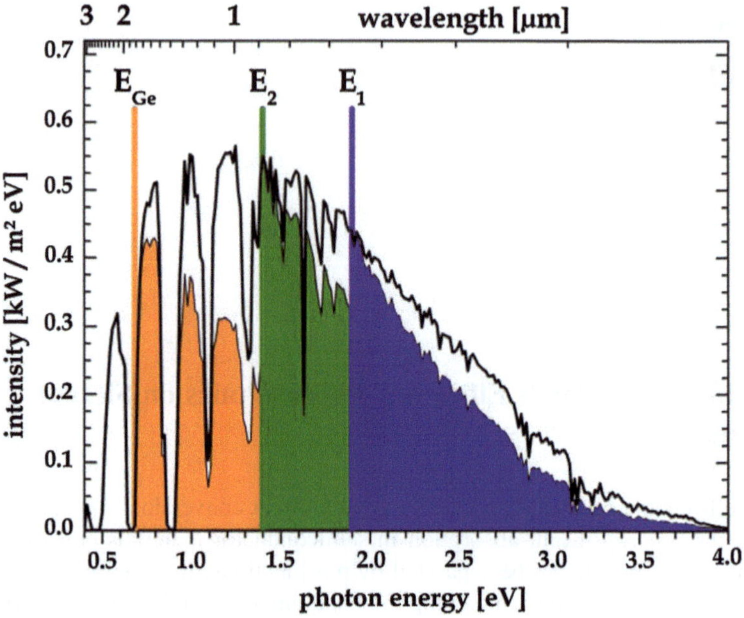

Fig. 1.1 Standard solar spectrum defined by ASTM G-173-03 for terrestrial irradiance considering an effective air mass of 1.5 atmospheres. *Colored lines* indicate the absorption edges of a typical semiconductor combination applied in current triple-junction solar cells, while the *colored area* shows an upper limit for the achievable electrical yield only considering transmission below the band gaps and thermalization of the charge carriers to the band edges

1.2 State-of-the-Art: III–V Semiconductor Heteroepitaxy on Si(100)

Beyond supplying economically and technologically advantageous quasi-substrates for multi-junction solar cells and other important applications such as lasers and LEDs, the technological interest in the superior electronic and optoelectronic properties of III–V semiconductors promotes the research in heteroepitaxy on silicon (Si) substrates for integration in established microelectronics. Hence, intense research for heteroepitaxial III–V integration on Si(100) substrates was started already in the early 1980s [3].

The plot of classical III–V and elemental semiconductor band gap energies over lattice constants shown in Fig. 1.2 summarizes potentials and challenges for the manufacture of epitaxial device structures. In general, the growth of ternary or higher order III–V compounds enables arranging and combining direct band gap energies according to the desired applications, but preserving highest epitaxial quality in a device structure usually requires maintaining the lattice constant of the substrate. Early III–V double-junction solar cells based on the growth of InGaP top cells after established homoepitaxial integration of GaAs bottom cells. Current

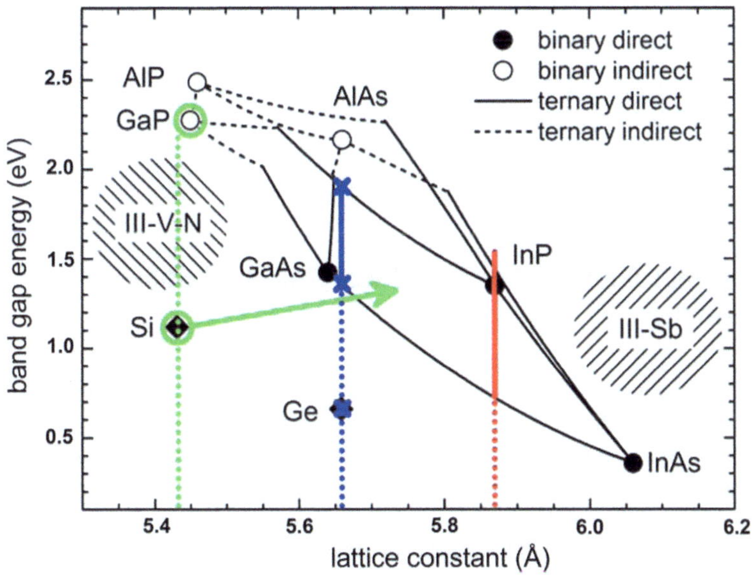

Fig. 1.2 Band gap energies of classical elemental and III–V compound semiconductors over lattice constants. *Solid* and *dashed black lines* indicate ternary III–V materials, the enclosed area is available as quarternary or higher order compounds. Admixture of antimony and nitrogen may extend the range of material combinations significantly. Lattice matched heterostructures such as multi-junction solar cells can be grown on gallium arsenide, germanium (marked *blue*) and InP (marked *red*), while heterostructures on silicon (marked *green*) either involve significant lattice mismatch or the application of dilute nitrides

high-efficiency triple-junction solar cells [2] combine this tandem configuration with Ge(100) substrates that are simultaneously used as low band gap bottom cell (Fig. 1.1). Due to the serial connection of the individual sub-cells, current matching is required to extract the highest output power. The integration of further sub-cells requires implementation of intermediate band gap energies in the range of 1.0 eV, not available on the lattice constants of GaAs and Ge. Hence, the integration of multi-junction solar cells on the lattice constant of InP represents a promising alternative [4], but the cost and availability of the substrate material is an issue. In contrast to Ge(100) substrate, the heteroepitaxial integration of classical III–V compound semiconductor on Si(100) substrates always requires a major change of the lattice constant, in particular to reach the composition regime of direct band gap materials.

Major challenges have to be met at the III–V/Si heterojunction [3]; several severe issues are related to crucial differences in general material parameters on the one hand and to the formation of the heterovalent (polar-on-non-polar) interface on the other hand. Compared to standard homoepitaxial III–V layers, heteroepitaxy on Si(100) has to face critical defect mechanisms including:

1. lattice mismatch (→ misfit dislocations)
2. different thermal expansion (→ strain after growth)
3. interdiffusion (→ undesired cross-doping)
4. polarity of the interface (→ violation of charge neutrality)
5. anti-phase disorder (→ anti-phase boundaries).

1.3 Outline: Gap/Si(100) Interfaces and Anti-phase Disorder

In contrast to other III–V compound semiconductors, gallium phosphide is almost lattice matched to silicon (Fig. 1.2). Below the critical layer thickness at which the accumulated strain enforces lattice relaxation by the formation of misfit dislocations, GaP/Si(100) heteroepitaxy results in pseudmorphic films which represent an ideal III–V/Si(100) model system for the investigation of issues related to the polar on non-polar interface in the absence of other defect mechanisms related to the differences in lattice constants and thermal expansion coefficients. Simultaneously, well-ordered heteroepitaxial GaP/Si(100) surfaces could potentially be used as quasi-substrates for the integration of III–V-based optoelectronic devices on Si(100) substrates either via graded buffer layers [5] or by the development of adequate dilute nitride III–V compounds lattice matched to Si(100) [6].

The schematic diagram of Fig. 1.3 illustrates the formation of anti-phase disorder—one of the most crucial defect mechanisms in III–V/Si(100) heteroepitaxy—at an idealized GaP/Si(100) interface. Although the crystal structures as well as the lattice parameters are very similar, zinc blende GaP exhibits a reduced symmetry to diamond cubic Si. While Si(100) surfaces can develop both monatomic or biatomic

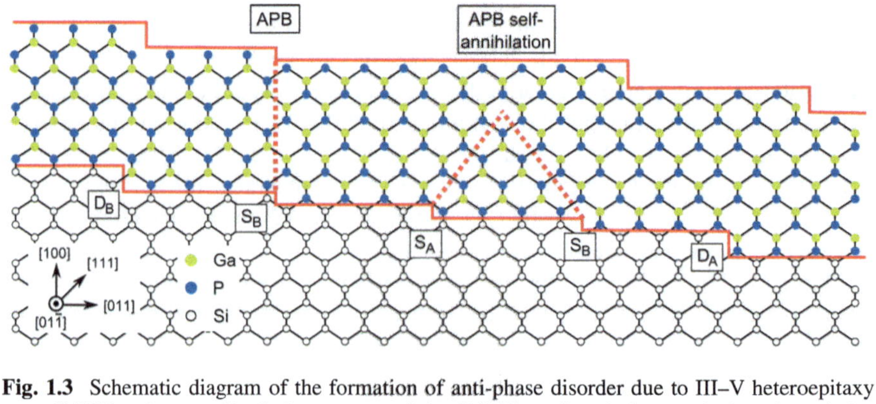

Fig. 1.3 Schematic diagram of the formation of anti-phase disorder due to III–V heteroepitaxy on a Si(100) substrate. Single-layer steps of the Si(100) surface induce anti-phase boundaries (APBs) which penetrate through the film and can be subject to self-annihilation, while double-layer steps on the substrate prevent the formation of APBs during nucleation

layer steps, GaP(100) always forms biatomic surface steps. During GaP nucleation, single-layer steps of the Si(100) substrate initiate anti-phase boundaries (APBs), which propagate with growth and separate anti-phase domains (APDs). For the fabrication of high-performance optoelectronic devices, a significant reduction of defect concentration is required and can only be achieved if anti-phase disorder is suppressed, either by double-layer step formation on the Si(100) substrate or by pair wise self-annihilation of APBs during growth of the GaP layer.

The present work reports about detailed interface investigations dedicated to the control of anti-phase disorder during early stages of heteroepitaxial GaP integration on Si(100) in a production-oriented MOVPE reactor. In an ambient determined by continuous process gas flow, the preparation and atomic structure of Si(100) surfaces differs significantly from standard UHV-based results (Chap. 3). Thorough evaluations of GaP(100) surface structures and the associated in situ RAS signatures (Chap. 4) enabled the development of a quantitative in situ probe of III–V/Si(100) anti-phase disorder (Chap. 5), which has been benchmarked with conventional ex situ characterization techniques as well as with different surface science techniques.

References

1. W. Shockley, H.J. Queisser, J. Appl. Phys. **32**, 510 (1961)
2. W. Guter, J. Schone, S.P. Philipps, M. Steiner, G. Siefer, A. Wekkeli, E. Welser, E. Oliva, A.W. Bett, F. Dimroth, Appl. Phys. Lett. **94**, 3 (2009)
3. S.F. Fang, K. Adomi, S. Iyer, H. Morkoc, H. Zabel, C. Choi, N. Otsuka, J. Appl. Phys. **68**, R31–R58 (1990)
4. H.J. Schimper, Z. Kollonitsch, K. Moller, U. Seidel, U. Bloeck, K. Schwarzburg, F. Willig, T. Hannappel, J. Cryst. Growth **287**, 642–646 (2006)
5. S. Verhagen, Master thesis, Radboud University Nijmegen, 2010
6. K. Volz, J. Koch, F. Hohnsdorf, B. Kunert, W. Stolz, J. Cryst. Growth **311**, 2418–2426 (2009)

Chapter 2
Experimental

2.1 Metal-Organic Vapor Phase Epitaxy

The application of specifically designed metal-organic (MO) precursor molecules enables epitaxial growth of various materials, in particular of compound semiconductors, by chemical vapor deposition (CVD) processes. The required chemical elements for the formation of epitaxial layers are provided at the substrate surface by thermal decomposition (pyrolysis) of metal-organics or metal hydrides. In contrast to molecular beam epitaxy (MBE), the growth process is a chemical reaction under process gas flow at moderate pressure, and not physical deposition in ultra-high vacuum (UHV). Due to advantages such as industrial scalability, reliable process control and established preparation of phosphorus containing III–V compounds, MOVPE has become the dominant process for the manufacture of laser diodes, multi-junction solar cells, and LEDs.

Most sample preparation processes reported in this work have been conducted in a modified commercial MOVPE reactor (Aixtron AIX 200), a traditional horizontal low-pressure reactor with an inner quartz liner and an outer quartz reactor. The graphite susceptor is heated by infrared lamps. Either H_2, purified in a palladium cell, or N_2, purified in a getter column, were utilized as the carrier gases at pressures between 50 and 950 mbar and at a total gas flow rate of 5.5 l/min. The applied precursor sources include tertiarybutylphosphine (TBP) and triethylgallium (TEGa) for GaP growth and dilute silane (SiH_4) for the deposition of silicon buffer layers. As a requisite for their subsequent transfer to UHV (Sect. 2.1.2), samples were clamped to a specifically designed molybdenum sample holder, which was fitted into an appropriately profiled niche engraved into the graphite susceptor. References [1, 2] describe the utilized MOVPE equipment in more detail.

However, the investigation of III–V heteroepitaxy on Si(100) presented in this work required several modifications of the MOVPE system which include:

1. An upgrade of the infrared heating system increased the maximum electrical input power from 6.0 to 9.6 kW provided the capability to reach process

H. Döscher, *GaP heteroepitaxy on Si(100)*,
Springer Theses, DOI: 10.1007/978-3-319-02880-4_2,

temperatures of up to 1,050 °C, which covers the relevant temperature range for the investigation of Si(100) surface preparation.
2. The installation of a nitrogen purged double-O-ring system provided enhanced protection of the sample preparation processes against extrinsic contaminations such as oxygen by diffusion from the ambient atmosphere.
3. An extension of the precursor supply system also included the installation gas source inlets enabling the use of gaseous group IV precursors for the growth of essential buffer layers on non-polar substrates such as Si(100).
4. The enhancement of optical in situ monitoring instrumentation (LayTec Epi-Curve TT and EpiRAS TT) also provided the capability for emissivity-corrected surface temperature measurement, which identified pressure dependent deviations between the crucial surface temperature of the samples and the thermocouple-controlled process temperature in the susceptor.

Several epitaxy processes were conducted in a different, but very similar MOVPE reactor system (Aixtron AIX 200-GFR), which is described by Ref. [3] in detail.

2.1.1 Substrates and Wet-Chemical Pretreatment of Si(100)

Our investigation of the VPE preparation of Si(100) surfaces started with different substrates misorientated by 0.1°, 2° and 6° in [011] direction, which were either covered by a thin, protective oxide layer [4] from wet-chemical pretreatment or just native oxides for comparison. In detail, the wet-chemical procedure involved cleaning in an ultrasonic bath (organic solvent), boiling in ammonium hydroxide (NH_4OH)-based solution prior to dilute HF-dip (3 % in deionized water), final boiling in hydrochloric acid (HCl) based solution as well as rinsing in deionized water between all steps as described by Ref. [4]. GaP homoepitaxy experiments were conducted for reference on sulphur-doped (2×10^{18} dopants/cm^3), exactly oriented GaP(100) substrates.

2.1.2 Essential Process Steps for Sample Preparation

The present work describes the reliable removal of protective oxide layers in a MOVPE reactor from Si(100) substrates by thermal annealing in hydrogen in detail (Sect. 3.2). Also, the growth of Si buffer layers via dilute SiH_4 at temperatures below 1,000 °C and the effect of subsequent cooling to the applied GaP growth temperatures were studied. X-ray photoelectron spectroscopy (Sect. 2.3.1) verified the absence of any contaminations besides hydrogen (not detectable) and minor arsenic adsorbates due to previous III–V deposition processes on the Si(100) surface after preparation. To avoid an impact of As on the Si(100) surface structures [5, 6], thorough conditioning of the MOVPE reactor reduced the As concentration to about the XPS detection limit (roughly 1 % of a monolayer in our instrument). Probably, most of the residual As adsorbed during the low pressure conditions required for the UHV-transfer, when the samples were already cooled well below the process temperature and no further reaction with the Si(100) surfaces was expected.

Heteroepitaxial GaP deposition started with pulsed low-temperature nucleation in the range of 400–450 °C consisting of alternating injections of the precursors TBP and TEGa. Subsequently, the growth temperatures for the actual GaP films on the Si(100) substrates were usually fixed in the range of 600 °C as typically also chosen for GaP homoepitaxy. Most GaP/Si(100) processes were stopped well before reaching the relaxation thickness of the GaP top layer, yielding pseudomorphic films with different thicknesses typically in the range of 15–100 nm.

Deoxidation of the GaP(100) and InP(100) substrates was achieved by annealing at 630 °C for 5 min using H_2 as carrier gas. Thermal decomposition of the III–V surfaces during annealing was prevented by TBP supply (at a partial pressure of about 20 Pa) in the MOVPE reactor, which stabilizes their P-rich termination. Afterwards, heteroepitaxial GaP/Si(100) and GaP(100) reference surfaces can be treated alike as verified (Sect. 5.3.1) and described in much detail (Sect. 4.2) by the present study. For the visualization of APD structures as height contrast (Sect. 5.1.1), a specific annealing procedure was applied to the samples after growth which involves heating to about 700 °C for a period of 10 min without or with low TBP supply.

2.1.3 Sample Transfer to Ultra High Vacuum

Subsequent to preparation in the MOVPE reactor, many samples were directly transferred to UHV utilizing a dedicated transfer mechanism to a mobile chamber [7] to provide unique contamination-free access for advanced surface characterization. On the one hand, MOVPE is the method of choice for manufacturing epitaxial III–V devices often involving heterointerfaces as their most critical components, on the other hand only very few surface sensitive instruments are available in process gas environments and the surfaces rarely remain stable during sample transport through air. Although RAS (Sect. 2.2.1) is applicable for the in situ identification of surface structures, there is no intuitive interpretation of RAS data. The assignment of characteristic spectra to specific surface reconstructions and the determination of the influence of other surface features on the shape and intensity of RAS peaks require extensive comparison with UHV-based surface science techniques such as XPS (Sect. 2.3.1), LEED (Sect. 2.3.2), STM (Sect. 2.3.3) and LEEM (Sect. 2.3.5). The principles of this benchmarking process, the MOVPE to UHV transfer concept and the utilized experimental equipment are described by Ref. [8] in more detail.

2.1.4 Access to Remote Instrumentation

While the experimental set-up of our standard surface characterization tools such as XPS and LEED was designed to match with the specific sample carrier design developed for MOVPE processing and subsequent UHV transfer, access to more sophisticated techniques usually requires adjustment of the sample carrier and longer transfer distances. In the case of the LEEM results reported in this work (Sect. 5.6), we demonstrated contamination-free UHV transfers to a remote instrument located at the TU Clausthal (in a distance of about 300 km from the HZB in Berlin) and sample exchange between incompatible carrier systems. In contrast to the compact molybdenum MOVPE sample holder, standard LEEM carriers are loaded in air and provide an integrated filament for radiant and electron impact heating to enable surface cleaning by thermal annealing in UHV. To connect both systems, we utilized an established adaptor concept to insert significantly smaller standard STM carrier plates in the MOVPE holder and designed a specific UHV lock scheme for the switch-over to the LEEM instrument. Particular modifications of the different sample carrier systems for transfer and the success of the subsequent LEEM measurements include the integration of a guide rail for the STM plates in the LEEM carriers and the implementation of sample-sized notches and retaining springs on the STM plates, which had to be manufactured completely out of MOVPE-compatible molybdenum.

A different concept for access to remote surface characterization techniques is to re-prepare the surface features of interest after transport through air and requires the capability to conduct adequate processing in the destination instrument. For the indirect characterization of Si(100) substrates after GaP deposition presented in this work, we realized the transport of specifically grown samples to a different MOVPE reactor capable for optical in situ characterization, where we succeeded in the re-preparation of the P-rich surface reconstruction required for further analysis.

2.2 In Situ Process Control

Reliable in situ control over epitaxial processes is of major interest for industrial manufacturing of optoelectronic devices. In particular, the stoichiometry of compound semiconductor surfaces is essentially influenced by the type of surface reconstruction and selective preparation is desirable for controlling crucial III–V heterointerfaces. In the case of MOVPE, the presence of carrier gases during processing limits the availability of in situ measurement techniques, in particular of surfaces sensitive ones. Simultaneously, the need for in situ optical surface characterization even increases due to certain surface structures, which are typical for MOVPE preparation, but hard to reproduce without the presence of a process gas such as the P-rich surface reconstructions of GaP(100) and InP(100). As shown in the present work by the example of Si(100), surface preparation in a common

MOVPE reactor can lead to fundamentally different results than expected according to the comprehensive knowledge obtained by UHV experiments over decades.

2.2.1 Reflectance Anisotropy Spectroscopy

Reflection anisotropy spectroscopy (RAS) is a very surface sensitive optical probe that is applicable for in situ observation in the MOVPE environment [9] and enables the identification of the atomic surface structures. RAS measures the difference Δr in the complex Fresnel reflection indices r along two perpendicular crystal axes x and y, normalized to their sum [10]:

$$\frac{\Delta r}{r} = 2\frac{r_x - r_y}{r_x + r_y} \tag{2.1}$$

Applied to cubic crystals without bulk anisotropy, the RAS signal provides access to the reduction of symmetry represented by the surface and its reconstruction. A detailed description of the RAS technique and a set-up similar to that used in this work can be found in the work of Aspnes et al. [11]. In the present study, most MOVPE preparation processes were monitored by a commercially available in situ RAS spectrometer (LayTec EpiRAS 200 or EpiRAS 200 TT). Besides the type of surface reconstruction, the impact of several other factors changes the shape and intensity of RA spectra. The influences by the sample temperature, the atomic order of the surface, the presence of adsorbates, the domain structure of the sample and the dimer orientation on the surface had to be considered separately for the interpretation of the RAS data presented in this work.

Specific RA spectra in the regime of ambient and growth temperatures were taken directly in the MOVPE reactor. In contrast, measurements at low temperature (−250 °C) were conducted in a UHV system utilizing a dedicated MOVPE to UHV sample transfer system which was also used to correlate RAS 'fingerprints' with the actual surface reconstructions identified in UHV [7] (and references therein). In specific in situ experiments the sample temperature was ramped up with 1.5 K/min at 50 mbar under pure hydrogen flow in order to monitor and observe transitions between different surface reconstructions in detail. Meanwhile, RA spectra were taken continuously (5 min per spectrum), and so the changes of the RA spectra and the temperature are represented in color-coded 2D plots over photon energy and temperature ('colorplot').

2.2.2 Curvature Sensor and Surface Temperature

We acquired an alternative optical in situ characterization technique (LayTec Epi-Curve TT), which allowed reliable surface temperature measurement by emissivity corrected pyrometry. Its application during Si(100) surface preparation processes was essential for the accurate interpretation of our results. The instrument and, in particular, its primary application for the control of strain induced bowing of substrates due to lattice mismatched graded buffer layers on GaP(100) and heteroepitaxially grown GaP/Si(100) quasi-substrates is described by Ref. [12] in detail.

2.2.3 *Mass Spectroscopy*

To obtain complementary in situ information about the vapor phase composition during the MOVPE processes, we installed a commercially available mass spectrometer (Hiden Analytical 301/3F) at the UHV compartment of our MOVPE system, which probes the process gas via throttled bypass line. Conclusive results about the decomposition products of protective oxide films during thermal annealing of Si(100) substrates (Sect. 3.2.3) are expected soon.

2.3 Surface Characterization in UHV

2.3.1 *X-ray Photoelectron Spectroscopy*

The identification of characteristic core level photoemission lines induced by incident X-ray radiation enables surface sensitive measurement of the chemical composition of samples. Since the exact core level binding energy of a species depends on its chemical coordination in the sample ('chemical shift'), accurate measurement of the energetic position of the observed photoemission peaks and detailed analysis of the observed line shape provides access to the chemical structure of the sample. For thorough evaluation of the Si(100) surfaces after preparation and identification of potential contaminations, an existing XPS system with a hemispherical electron analyzer (Specs Phoibos 100) was upgraded by installation of a monochromated $Al_{K\alpha}$ X-ray source (Specs Focus 500). The influence of band bending due to different termination of the Si(100) surfaces was corrected by matching the energetic positions of the major elemental silicon photoemission signals between different samples. Detailed chemical analysis of the XPS spectra was supported by sophisticated deconvolution procedures enabled by commercial software (CASA XPS) [13].

2.3.2 *Low-Energy Electron Diffraction*

The lateral symmetry of surface structures and particularly surface reconstructions can be visualized by the diffraction induced by a normal-incidence electron beam typically using low kinetic energies in the range of 40–120 eV. For application of this standard surface characterization technique, we used a commercial set-up (Specs ErLEED 100-A) with reverse view LEED optics and recorded the diffraction pattern employing a digital video camera.

2.3.3 Scanning Tunneling Microscopy

The lateral dependency of the tunneling current between a sufficiently conductive sample and a sharp metal tip can be exploited to obtain atomically resolved images of the surface and study local surface structure by scanning. Our STM measurements were carried out with a commercial SPECS 150 Aarhus STM ($<5\times10^{-10}$ mbar) using tungsten tips prepared by electrochemical polishing in potassium hydroxide solution with direct current. To avoid frequent tip exchange, in situ tip conditioning was performed by argon ion sputtering. All images were recorded in constant-current mode with tunneling currents on the order of 100 pA; the bias voltage values quoted in the following refer to the sample voltage with respect to the grounded STM tip. Si(100) measurements were conducted with positive sample bias of about 0.8–1.3 V, while obtaining contrast on the GaP(100) surface required high, negative sample bias at around −3.5 V, in agreement with previous studies [14–16]. Experimental details about supplementary STM measurements of the P-rich InP(100) surfaces with an Omicron STM in UHV ($< 5\times10^{-10}$ mbar) were as previously published [17].

2.3.4 Fourier-Transform Infrared Spectroscopy

For the characterization of hydrogen bonds on the samples, we used a commercially available FTIR spectrometer (Bruker IFS 66v/s). Surface sensitive measurements required UHV conditions and parallelepiped shaped MOCVD samples to obtain an attenuated total reflection (ATR) configuration. Based on the comparison with spectra of untreated, still oxidized sample, we used a linear background as a reference for the frequency range relevant for Si–H stretch modes [18]. Details about the components and the specific design of our set-up can be found in Ref. [19].

2.3.5 Low-Energy Electron Microscopy

The use of electrons with low primary energy (0–30 eV) for microscopic investigations ensures high surface sensitivity [20]. While bright-field contrast depends on the electron reflection coefficient which is related to the band structure and the topography of the surface, dark-field contrast is determined by the selected diffraction spots of the LEED pattern which is characteristic for the surface reconstruction mechanism it is assigned to. LEEM is operated with a parallel electron beam for simultaneous illumination of the whole field of view. An aperture in the diffraction image plane allows the selection of specific diffraction spots of the observed LEED pattern for the generation of magnified surface images. In the present study we used two different contrast mechanisms for surface imaging of the heteroepitaxial GaP/Si(100) film: bright-field imaging via specular reflection using the (0,0) spot [20], and dark-field imaging employing higher-order diffraction spots [21] for the discrimination of the APDs on the surface. Further details about applicable imaging techniques and the utilized instrumentation are described in Ref. [22].

2.4 Ex Situ Characterization

2.4.1 Atomic Force Microscopy

Similar to STM, the mechanical interaction of a sharp tip with a sample surface also can be exploited to obtain microscopically resolved images by scanning. In the present study, tapping mode atomic force microscopy (AFM) was carried out ex situ for investigation of the μm scale surface morphology of Si(100) and GaP/Si(100) samples. Due to the application of a specific annealing technique [23], also APBs could be visualized as height contrast by AFM (Sect. 5.1.1). Since the obtained height modulation at APBs was large compared to the mono-stepped GaP surface, the two different phases could easily be distinguished and highlighted by color-coding for determination of their local surface fractions. All AFM measurement were conducted in air using two commercial instruments (both of them Digital Instruments Nanoscope Multimode IIIa) located in different laboratories.

2.4.2 Transmission Electron Microscopy

The transmission of high-energy electron beam through thin, specifically prepared specimens can be utilized to obtain microscopic resolution, and several imaging techniques exist to provide different types of contrast. Besides conventional high resolution imaging, we applied specific dark-field imaging modes for the detection of GaP/Si(100) anti-phase disorder (Sect. 5.1.1). The underlying contrast mechanisms are described in Ref. [3] in detail. For subsequent TEM characterization, {011} cross sectional as well as (100) plan view specimens were prepared conventionally by mechanically cutting, polishing, and using Ar-ion milling as the last preparation step. The TEM images shown in this study have been measured with two different instruments, a Philips CM12 microscope with an accelerating voltage of 120 kV and super twin lens modification and a Jeol JEM 3010 microscope, working at an acceleration voltage of 300 keV.

2.4.3 General Material Characterization

X-ray diffraction (XRD) is versatile tool for the characterization of epitaxial layers. In the present work, a high resolution XRD instrument (Panalytical X'Pert Pro MRD) was used for complementary layer thickness determination and for determination of lattice relaxation in heteroepitaxial GaP/Si(100), when growth exceeded the critical layer thickness. Details about the utilized set-up and the experimental techniques are provided by Ref. [12]. For the quantitative analysis of bulk compositions, in particular regarding minor impurity or dopand concentrations, samples were send to a company (RTG Mikroanalyse GmbH Berlin), which offers commercial secondary ion mass spectroscopy (SIMS) measurements.

References

1. U. Seidel, PhD thesis, Humboldt-Universität zu Berlin, 2007
2. K. Möller, PhD thesis, Universität Duisburg-Essen, 2007
3. I. Nemeth, PhD thesis, Philipps-Universität Marburg, 2008
4. A. Ishizaka, Y. Shiraki, J. Electrochem. Soc. **133**, 666–671 (1986)
5. T. Fox-Bork, PhD thesis, Humboldt-Universität zu Berlin, 2007
6. T. Hannappel, W.E. McMahon, J.M. Olson, J. Cryst. Growth **272**, 24–29 (2004)
7. T. Hannappel, S. Visbeck, L. Toben, F. Willig, Rev. Sci. Instrum. **75**, 1297–1304 (2004)
8. T. Hannappel, Habilitation, Freie Universität Berlin, 2005
9. J.T. Zettler, Prog. Cryst. Growth Charact. Mater. **35**, 27–98 (1997)
10. P. Weightman, D.S. Martin, R.J. Cole, T. Farrell, Rep. Prog. Phys. **68**, 1251–1341 (2005)
11. D.E. Aspnes, J.P. Harbison, A.A. Studna, L.T. Florez, J. Vac. Sci. Technol. A-Vac. Surf. Films **6**, 1327–1332 (1988)
12. S. Verhagen, Master thesis, Radboud University Nijmegen, 2010
13. N. Fairley, CasaXPS Version 2.3.13Dev22: Software Package for Data Processing. http://www.casaxps.com
14. L. Töben, T. Hannappel, K. Moller, H.J. Crawack, C. Pettenkofer, F. Willig, Surf. Sci. **494**, L755–L760 (2001)
15. N. Kadotani, M. Shimomura, Y. Fukuda, Phys. Rev. B **70**, 5 (2004)
16. P.H. Hahn, W.G. Schmidt, F. Bechstedt, O. Pulci, R. Del Sole, Phys. Rev. B **68**, 033311 (2003)
17. P. Vogt, T. Hannappel, S. Visbeck, K. Knorr, N. Esser, W. Richter, Phys. Rev. B **60**, R5117–R5120 (1999)
18. Y.J. Chabal, G.S. Higashi, S.B. Christman, Phys. Rev. B **28**, 4472–4479 (1983)
19. A. Dobrich, Diploma thesis, Freie Universität Berlin, 2009
20. E. Bauer, Rep. Prog. Phys. **57**, 895–938 (1994)
21. H. Marbach, G. Lilienkamp, H. Wei, S. Günther, Y. Suchorski, R. Imbihl, Phys. Chem. Chem. Phys **5**, 2730–2735 (2003)
22. B. Borkenhagen, Diploma thesis, TU Clausthal, 2006
23. I. Nemeth, B. Kunert, W. Stolz, K. Volz, J. Cryst. Growth **310**, 1595–1601 (2008)

Chapter 3
Si(100) Surfaces in Chemical Vapor Environments

3.1 State-of-the-Art: Si(100) and its Atomic Surface Structure

The epitaxial integration of various compound semiconductors such as GaN, SiC or classical III–V materials on silicon (Si) substrates represents a technological challenge of increasing importance. The aim to stand the current pace of advancement in Si-based microelectronics particularly demands the direct implementation of optoelectronical components. Simultaneously, the fast progress of the III–V based optoelectronic industry reinforces the need for advantageous alternative substrates as stable, inexpensive and readily available as the established Si technology already is today. However, the obstacles of compound semiconductor growth on Si(100) [1] include several issues associated with the heterovalent (polar-on-non-polar) interface. Prior to heteroepitaxy, the atomic surface structure of the substrate already determines the initiation of undesirable anti-phase disorder [2] according to the distribution and height of steps (Fig. 1.3). In principle, an elaborate Si(100) surface preparation may not only avoid contamination of the crucial heterointerface but, when double-layer steps can be achieved, may also prevent the formation of anti-phase domains (APDs) [3].

Intense III–V/Si investigations have been started in the 1980s mainly motivated by the potential integration of GaAs elements with the established microelectronic technology based on Si(100) substrates [2, 4]. The mature molecular beam epitaxy (MBE) technique for GaAs heteroepitaxy [5, 6] enables implementation of the comprehensive experience with the Si(100) surfaces in ultra high vacuum (UHV). There, the preparation of clean Si(100) induces the formation of Si-dimers [7] which compensate only half of the dangling bonds and basically induce a (2×1) symmetry. Typically, single-layer surface steps emerge and separate terraces with perpendicular dimer orientation [7, 8] resulting in the characteristic two-domain $(2 \times 1)/(1 \times 2)$ surface reconstruction of Si(100). Driven by a potential reduction of surface energy according to Ref. [9], the development of double-layer steps appears favorable (Fig. 3.1), but only occurs predominantly on vicinal surfaces with high step density [10–12].

H. Döscher, *GaP heteroepitaxy on Si(100)*,
Springer Theses, DOI: 10.1007/978-3-319-02880-4_3,

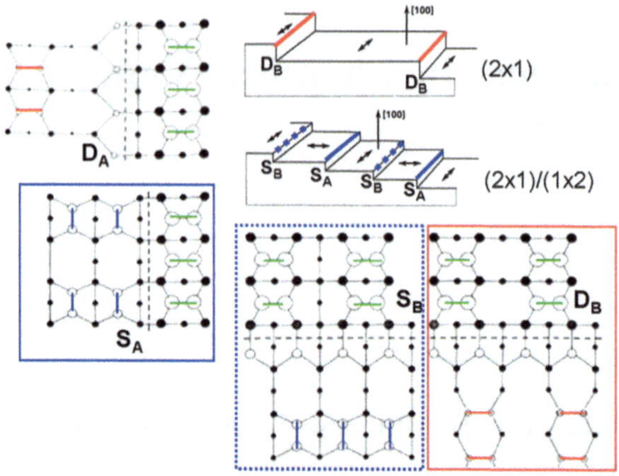

Fig. 3.1 Schematic sketch of the atomic surface structure of Si(100) defining the type of step by the dimer orientation relative to the step edge according to Ref. [9]. Open (*filled*) *circles* represent Si atoms of the surface (sub-surface layers), their size indicates relative height. Dimers on the upper terrace are marked by *green bars*, those one and two atomic layers below by *blue* and *red bars*, respectively. Since dimer orientation switches at single-layer steps, both types have to coexist independent of differences in step formation energy, unless the formation double-layer steps of which type B is considered energetically favorable enables a single-domain surface reconstruction (After [9])

However, clean Si(100) surfaces are quite reactive surface and susceptible to contamination [13], including immediate, so-called native oxidation [14] in air as well as other strong chemical bonds with carbon (SiC), nitrogen (Si_3N_4), and other elements. Wet-chemical processing with hydrofluoric acid (HF) dissolves SiO_2 and results in a rather stable hydrogen termination of Si(100) substrates, but the recontamination of the surface is only suppressed to some degree. Instead, common wet-chemical treatment strategies for Si(100) finish with well-defined reoxidation ('Shiraki' method) [15] to form a thin, protective SiO_2 layer intended for thermal removal directly in the epitaxy system.

In the expanding optoelectronic industry, metal–organic vapor phase epitaxy (MOVPE) is the prevailing manufacturing technology. Besides reliable process control and industrial scalability, the more established preparation of nitride [16] and phosphide [17] based compound semiconductors is considered a key advantage. The demand for large-scale production of sophisticated optoelectronic devices such as LEDs, lasers, or multi-junction solar cells increases the desire for improved heteroepitaxial III–V growth on Si(100) via MOVPE to benefit from more abundant alternative substrates, larger wafer sizes, improved mechanical properties etc.

Purified hydrogen is the common standard for MOVPE process gas, but there are only few studies of the Si(100) surface preparation and its peculiarities in the vapor phase, since many surface sensitive characterization techniques require UHV and are not directly available. Various III–V/Si(100) investigations with MOVPE report the growth of heteroepitaxial GaAs [18], InP [19], or GaP [20] based structures on differently prepared substrates. However, the impact of typical MOVPE process gas conditions on the preparation and the atomic surface structure of Si(100) substrates is less well-characterized. Molecular hydrogen can dissociate and adsorb on Si(100) surfaces at typical process temperatures [21, 22]. Several UHV based studies [23–26] address the influence of hydrogen on Si(100) and particularly analyze its impact on the atomic surface structure as depicted in Fig. 3.2.

For subsequent III–V integration, the literature emphasizes two crucial requisites on Si(100) substrates:

1. Purification of the highly reactive surface, which is particularly susceptible to native oxidation, and
2. Suppression of APD formation by control over the atomic surface structure ideally arranged in double-layer steps.

In analogy to the UHV preparation of Si(100), the formation of well-established protective oxide layers [15] can reduce the substrate cleaning in the MOVPE system to thermal SiO_2 removal immediately before starting the subsequent heteroepitaxial deposition. However, the presence of the process gas influences the deoxidation process significantly: several CVD studies [27–30] agree that oxide removal in hydrogen requires heating to about 950 °C, whereas similarly treated Si(100) substrates can be cleaned by annealing at 800 °C or even below in UHV [15]. Physical and chemical influences on the deoxidation reaction can be discussed [31], but also major influences on the atomic structure of the cleaned Si(100) surface have to be expected: in UHV-based experiments, the saturation of the dangling bonds with hydrogen [26] changes the experimentally observed step structures [25] significantly.

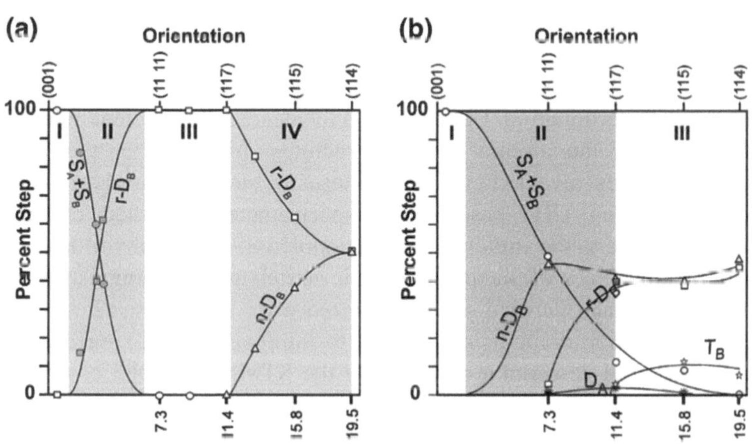

Fig. 3.2 Experimental STM study of the distribution of step types on clean (**a**) and monohydride terminated (**b**) Si(100) over misorientation angle as reported by Laracuente and Whitman [25]. While double-layer steps prevail on clean vicinal Si(100) beginning with a few degrees off-cut, a significant share of single-layer steps remains present over a wide angular range (From [25])

Only a few studies consider Si(100) surface structures prepared in CVD environments, in most cases rather indirectly derived from the analysis of subsequent epitaxial layers [31, 32]. Reference [33] describes the surface structure of group V terminated Si(100) as obtained by annealing in phosphine and arsine, respectively. Reference [30] analyzes the thermal removal of a protective SiO_2 layer and shows by ex situ atomic force microscopy (AFM) that further annealing in hydrogen reduces the surface roughness of Si(100). Also employing ex situ AFM, Ref. [27] reports on the surface structure of slightly off-cut Si(100) substrates after CVD preparation for certain Si(100) substrates with slight misorientation in [011] direction: In contrast to Si(100) surface investigations in UHV [7, 10, 12, 25], a tendency for double-layer step formation after the growth of a homoepitaxial buffer layer was observed.

The present study reports on a reliable process for Si(100) preparation in a CVD environment and its systematic investigation by means of XPS. A dedicated sample transfer system to UHV [34] enabled us to preserve the Si(100) surfaces in the state prepared in our MOVPE reactor. We verified the capability to obtain clean Si(100) in a MOCVD reactor by XPS and observed the removal of the protective oxide layer when varying basic process parameters such as surface temperature and annealing time [35]. Enhanced surface sensitivity obtained by smaller photoelectron take-off angle enabled us to analyze residual SiO_2 layers in detail and to identify partially oxidized Si states associated with their interface to the Si(100) substrate by deconvolution of the XPS signals. The determination of the average residual oxide thickness provided a characteristic measure for the progress of oxide removal. According to the rather abrupt deoxidation behavior, we defined a critical temperature for oxide removal and investigated its dependence on the chemical vapor ambient, considering pressure and precursor injection as well as nitrogen and argon as alternative process gases [36]. After processing in hydrogen, complementary atomic surface structure investigations by LEED and STM usually resolved a two-domain $(2 \times 1)/(1 \times 2)$ reconstruction due to dimer formation and single-layer steps on the surface typical of clean Si(100) as well as of a monohydride termination. Depending on the degree of misorientation, detailed characterization of the atomic surface structures directly after thermal oxide removal in hydrogen revealed complex patterns of steps and terraces [37]. ATR measurements in an UHV-based FTIR spectrometer identified characteristic absorption lines due to the stretch modes of coupled Si–H monohydrides [38]. The polarization dependence of the antisymmetric stretch mode distinguished different dimer orientations and verified a clear preference for one of the $(2 \times 1)/(1 \times 2)$ surface reconstruction domains subsequent to improved Si(100) surface preparation. Tip-induced H-desorption observed in the STM proved the complete saturation of dangling bonds after VPE-preparation. During annealing in the MOVPE reactor at elevated temperature, in situ reflectance anisotropy spectroscopy showed the absence of Si–H bonds. Close inspection of the atomic surface structure of differently misoriented Si(100) surfaces by STM after improved VPE processes showed unconventional, D_A-like steps [39], which are considered to be energetically unfavorable (Fig. 3.1). For slight off-cut substrates, high resolution images revealed that there was an intermediate S_B step with associated dimer rows, only a few dimers long, in most places. Also on the 2° misoriented samples, we observed D_A-like steps, but only when cooling at the end of the process was slowed down. Here, also intermediate S_B steps were visible, but in a few places true D_A steps seemed to prevail.

3.2 Thermal Oxide Removal in Hydrogen Under Standard Process Conditions

Since Si(100) surfaces react with virtually any organic or inorganic contamination to form undesirable impurities, we used the well-defined reoxidation of the substrate by a subsequent wet-chemical step [15] to form a protective layer as starting point of our investigations. Basically, this well-established procedure [3, 27, 28, 40] simplified the CVD-preparation of a pure Si(100) surface, as required for subsequent III–V deposition, to the removal of the protective SiO_2 layer by thermal annealing directly in the MOVPE system.

3.2.1 Verification of Thermal Oxide Removal by XPS

In contrast to previous MOVPE studies, we were able to investigate the chemical composition of the substrate surfaces directly after annealing in the CVD environment by means of XPS [34] to analyze the remaining contamination. Actually, a process consisting of 30 min annealing at 950 °C surface temperature under flow of purified hydrogen at nearly atmospheric pressure (950 mbar) eliminated the photoemission signals due to oxygen completely [35], while these were still present after an analogous process in which the sample was annealed at only 900 °C.

By comparison of the XP spectra in the range of the O1s and Si2p photoemission lines (Fig. 3.3), we verified an intact stoichiometric SiO_2 layer for the annealing procedure at lower temperature (thick blue spectra) and a completely deoxidized Si(100) surface for the one at higher temperature (thin black spectra). The O1s binding energy observed at about 533.1 eV complies well with the chemical shift of +2.1 eV compared to the elemental O1s line at 531 eV, as expected for oxygen bound in SiO_2 [41, 42]. In the case of the deoxidized sample (thin black spectra), the absence of the O1s signal agreed with an increased intensity of the entire Si2p photoemission line around the binding energy expected for elemental silicon. In detail, we observed a peak at about 99.69 eV and a clearly resolved shoulder at about 100.31 eV. The splitting of the peak positions of approximately 0.62 eV and the intensity distribution agreed well with the separation of $Si2p_{3/2}$ and $Si2p_{1/2}$ due to spin orbit coupling [43].

On the sample which was still covered with protective SiO_2 (thick blue spectra), the measured intensity of the elemental Si2p line decreased according to the

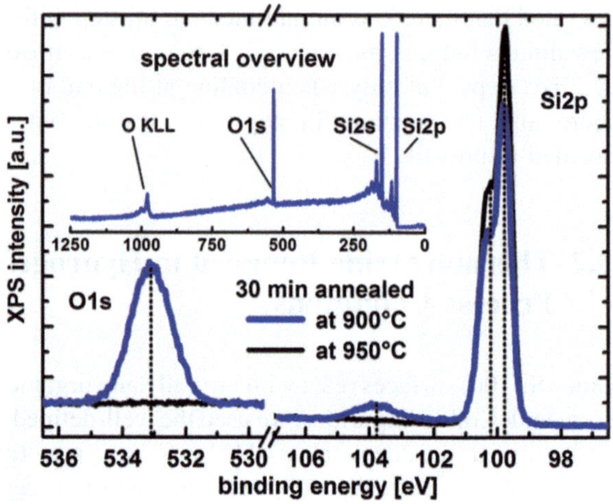

Fig. 3.3 O1s and Si2p XPS lines measured at two Si(100) samples after annealing in the MOCVD reactor under pure H_2 flow for 30 min at 900 °C (*blue*, *thick*) and at 950 °C (*black*, *thin*) surface temperature, respectively and an XPS overview of the non-deoxidized sample (*inset*) (From [35])

thickness of the SiO_2 layer, which was below the mean free path of the excited electrons. The Si2p photoemission from the SiO_2 layer itself formed an additional peak structure at about 103.7 eV (thick blue spectrum) corresponding to the higher oxidation state of Si in the oxidized film [44–46].

3.2.2 Progression of the Oxide Removal in Hydrogen Process Gas

According to our comparative experiments, we determined 950 °C as the critical temperature for the reliable removal of protective SiO_2 layers [15] from Si(100) substrates based on 30 min annealing in H_2 at nearly atmospheric pressure in a commercial MOCVD reactor. Since an analogous process at 925 °C surface temperature led to no significant reduction of the XPS signals associated to the silicon oxide (not shown here), we supposed either a very rigid thermal activation of the oxide removal or a highly non-linear progression of the deoxidation reaction. Distinct annealing experiments in the same process gas environment and at an identical temperature of 950 °C but with reduced annealing times led to conclusive XPS results shown in Fig. 3.4.

Even though Si(100) substrates were deoxidized reproducibly within 30 min H_2 annealing at 950 °C (Fig. 3.4, grey line), the interruption of this process after only 10 min annealing (Fig. 3.4, thick blue line) resulted in no significant reduction of the XPS intensities attributed to oxygen or oxidized silicon compared to insufficient oxide removal experiments conducted at lower temperatures.

In contrast to all other samples, which were either completely free of oxygen or covered with a laterally homogenous SiO_2 layer, XPS measurements subsequent to process interruption after 20 min H_2 annealing at laterally varied sample positions

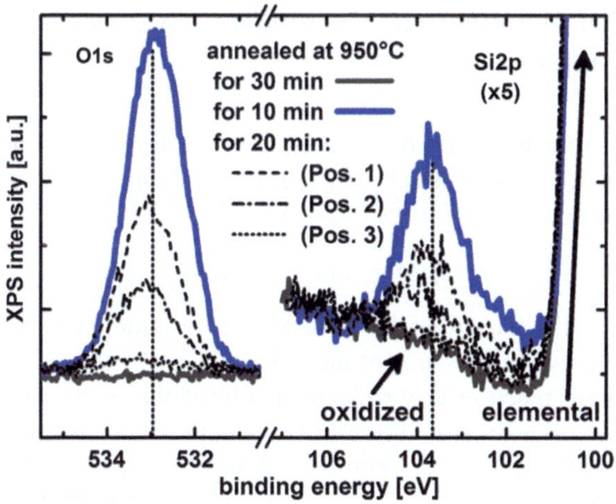

Fig. 3.4 O1s and Si2p XPS lines measured at Si(100) substrates after annealing in the MOCVD reactor under pure H_2 flow at 950 °C for 10 min (*thick blue line*), 20 min (*broken black lines*), and 30 min (*grey line*), respectively. The Si2p signal was amplified by a factor of 5 for better visibility of the oxidized share, while the main elemental line due to the substrate exceeds the applied scaling. The three broken black lines reflect very different residual oxide quantities coexistent on a single Si(100) substrate after the 20 min process at different, laterally distributed measurement positions (From [35])

(Fig. 3.4, broken black lines) actually led to significant differences in the residual oxide coverage of the Si(100) surface. Apparently, this experiment addressed a critical stage of the oxide removal from the surface and indicated a highly non-linear progression of the underlying deoxidation reaction. So far, we could not rule out completely that inhomogeneities regarding the initial oxide thickness or the lateral surface temperature distribution during annealing also contributed to the experimental result.

In any case, our findings showed a very abrupt deoxidation behavior including a high potential for lateral variations consistent with a complex SiO_2 removal mechanism such as void formation and growth, which has been established for Si(100) preparation in UHV [47–51]. However, the wet-chemical protective SiO_2 layer [15] we prepared on the Si(100) surface prior to introduction into the growth environment can be thermally removed by annealing at temperatures less than 800 °C in UHV [15], while in agreement with most MOVPE studies [27, 28, 52] we required roughly about 950 °C under H_2 flow. The progress of oxide removal in hydrogen has also been monitored in situ for different annealing temperatures employing RAS. Details about the in situ RAS observation of the deoxidation, hydrogen termination, domain distribution and step arrangement of Si(100) surfaces with different misorientations (0.1°, 2°, 6°) are in preparation and will be published elsewhere [53].

3.2.3 Potential Deoxidation Mechanisms in a Process Gas Environment

In UHV, the formation and desorption of volatile silicon monoxide (SiO) according to Eq. 3.1 is the generally accepted deoxidation mechanism for thermal surface preparation of Si(100) substrates [54]:

$$SiO_2(s) + Si(s) \rightarrow 2SiO(g) \uparrow \tag{3.1}$$

The process is limited either by poor SiO desorption from the SiO_2/Si(100) interface hindered by the presence of the oxide film or by the lack of elemental Si for the deoxidation reaction at the vacuum-side of the oxide film. However, as soon as the statistical formation of initial voids in the SiO_2 layer [50] exposes first parts of the Si(100) surface, the lateral mobility of Si atoms is locally enhanced there [48, 51]. Their diffusion to the edges of the voids largely promotes the deoxidation reaction according to Eq. 3.1; consequently, the growth of voids is an integral part of the deoxidation model for Si(100) surfaces in UHV [47, 49].

Previously, Yamazaki et al. [31] observed a need for higher annealing temperatures for oxide removal in hydrogen determined process gas environments compared to UHV and discussed physical limitations of the SiO desorption from the surface by kinetic restrictions due to the process gas flow for an explanation. However, Aoyama et al. [21] showed that at least a partial monohydride termination of (2 × 1)/(1 × 2) reconstructed Si(100) [55] occurs even after annealing at moderate sample temperatures and process gas pressures due to the dissociation of molecular hydrogen from the process gas at the surface, principally according the following reaction path:

$$-Si - Si - (s) + H_2(g) \rightarrow H - Si - Si - H(s) \tag{3.2}$$

The saturation of the dangling bonds by hydrogen potentially passivated the initially deoxidized parts of the Si(100)—largely restraining the diffusion of Si atoms. Hence the effect could disable the driving mechanism for the void growth mechanism according to Eq. 3.1 and provide an alternative explanation for the increase of the critical surface temperature for oxide removal in hydrogen compared to the UHV deoxidation of Si(100).

At the higher temperature for oxide removal in hydrogen, also an active role of the process gas in the deoxidation reactions such as the formation and desorption of water may have to be taken into consideration:

$$SiO_2(s) + H_2(g) \rightarrow H_2O(g) \uparrow + SiO(g) \uparrow \tag{3.3}$$

Although this reaction appears thermodynamically unfavorable at first, the dissociation of hydrogen molecules at incompletely saturated Si-dimers or continuous hydrogen desorption from the monohydride termination according to Eqs. 3.4a and b, respectively, might supply highly reactive hydrogen species as reactants to replace molecular hydrogen in Eq. 3.3:

$$-\text{Si}-\text{Si}-(\text{s})+\text{H}_2(\text{g})\rightarrow \text{H}-\text{Si}-\text{Si}(-\text{s})+\text{H}^*(\text{g}) \tag{3.4a}$$

$$\text{H}-\text{Si}-\text{Si}-\text{H}(\text{s})\rightarrow \text{H}-\text{Si}-\text{Si}(-\text{s})+\text{H}^*(\text{g}) \tag{3.4b}$$

If the actual deoxidation process in hydrogen involves this type of reaction path, the progress of oxide removal might be non-linearly promoted by the presence of already deoxidized parts of the Si(100) surface, in a certain analogy to the void growth processes observed in UHV [47–49].

To obtain conclusive results regarding the actual deoxidation mechanism in the CVD environment we plan several experiments involving alternative process gases and process gas compositions for Si(100) surface preparation as well as in situ process analysis by mass spectroscopy for detection of the chemical products of the deoxidation reaction [54]. Indications for an active role of hydrogen might be obtained by Fourier-transform infrared spectroscopy [21], optical in situ characterization [56], or even STM [57]. If laterally inhomogeneous oxide removal processes such as void formation in UHV also determine the deoxidation in the CVD environment, the transfer of relevant samples to surface sensitive microscopic instruments such as scanning Auger [48] or low-energy electron [47] techniques could provide direct evidence.

3.2.4 Surface Structure After Thermal Oxide Removal in a CVD Reactor

The previous sections showed that our process resulted in clean Si(100) surfaces after annealing under hydrogen at temperatures above 950 °C for more than about 20 min. In order to investigate the symmetry and atomic structure of the resulting surfaces, we carried out LEED and STM measurements as shown in Fig. 3.5 for a typical surface.

Figure 3.5a shows a typical LEED image of a surface prepared by our H_2 annealing process. The image reveals a (2 × 1)/(1 × 2) surface reconstruction, i.e. the surface is characterized by a unit cell which extends over the space of two Si atoms of the unreconstructed surface and which exists in both [011] and $[0\bar{1}1]$ directions, i.e. two domains exist which are rotated by 90° with respect to each other. The Si(100) surface is known to form Si–Si dimers, both at the clean surface as well as at the monohydride-covered surface. At the clean surface the dimers are buckled [58], but they perform a flip-flop motion leading to a symmetric appearance of the dimers [59]. Hydrogen exposure of this surface might lead to the formation of hydrides including symmetric monohydride H–Si–Si–H dimers [55]. Both, the clean reconstructed Si(100) as well as monohydride terminated surface are consistent with the observed LEED images. The presence of two orthogonal domains indicates that the surface is characterized by monolayer steps, where the orientation of the dimers changes by 90°. In Fig. 3.5b we present an empty state STM image of a small section of the surface providing atomic-scale evidence of a

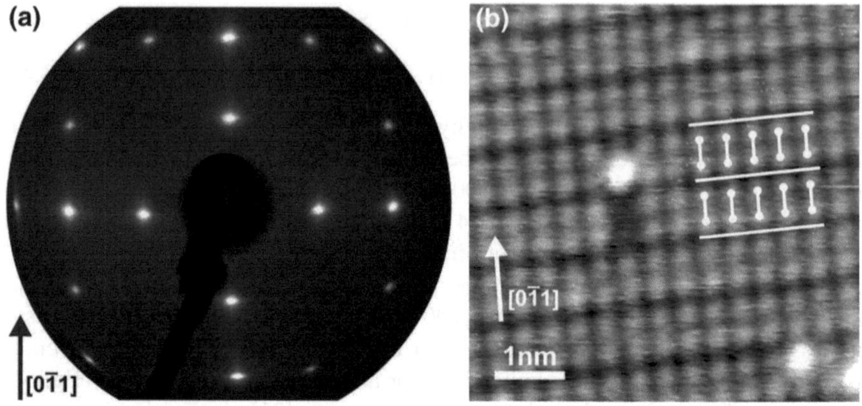

Fig. 3.5 **a** LEED image of a deoxidized Si(100) surface at an incident beam energy of 70 eV exhibiting half-order reflexes due to a two-domain (2 × 1)/(1 × 2) reconstruction. **b** Empty state STM image of a deoxidized Si(100) surface (V_{Sample} = 1.3 V, no filtering applied). Rows of individual dimers, which constitute a (2 × 1) surface unit cell, are clearly resolved (After [35])

well-ordered Si(100) surface structure. Individual protrusions are clearly visible which can be assigned to the silicon atoms at the surface. These appear as pairs in the image, forming rows separated by trenches, which are deeper than the gaps separating the atoms within the pairs and those separating the pairs within the rows. We identified the pairs of protrusions as Si–Si dimers, but similar to the LEED results, the STM image does not directly allow us to determine whether the surface is hydrogen terminated, since the appearance of the monohydride surface and that of the clean surface (depending on the applied bias voltage and shape of the tip) can be very similar [60]. There are several defects visible in the image: in one location a dimer appears to be missing, and additionally there are three bright protrusions. These may be dangling bonds at an otherwise hydrogen-terminated surface, impurity atoms at the site of silicon atoms, or individual adatoms.

3.3 Variation of the Process Environment and Alternative Carrier Gases

Thermal oxide removal from Si(100) substrates directly in the epitaxy system has been identified as initial process step for heteroepitaxial III–V integration, since the reproducible preparation of a protective SiO_2 layer as final wet-chemical pre-treatment [15] prevents uncontrolled reoxidation and critical carbon contaminations in air. In a recent study [35], we verified thermal oxide removal from Si(100) in the hydrogen ambient of a MOVPE reactor using XPS under normal photoemission.

3.3.1 *Residual Oxide Films After Hydrogen Annealing Observed*

In order to confirm the complete deoxidation of Si(100) substrates, we performed XPS at reduced photoelectron take-off angles $\alpha < 90°$ to further increase the surface sensitivity of the measurement. At a take-off angle of $\alpha = 30°$, we expected a relative gain of the signal contribution due to the topmost layers by a factor of 2 due to geometrical reasons. Consistent with normal emission measurements [35], we did not find any trace of oxygen in the XPS spectra after a 30 min annealing at a surface temperature of 950 °C under flow of purified hydrogen at a pressure of 950 mbar (Fig. 3.6), while clear signals were observed after an analogous process in which the sample was annealed at only 900 °C.

Figure 3.6 displays a comparison of the characteristic O1s and Si2p photoemission lines measured at both samples under a take-off angle of 30°: while we observed only the Si2p signature of pure Si(100) at about 99.7 eV after the 950 °C process (thin black spectra), also an intense O1 s line at about 533.1 as well as a secondary Si2p peak at about 103.7 eV were still measured after the 900 °C process (thick blue spectra). The chemical shifts associated with both peak positions matched with the binding energies expected for stoichiometric SiO_2 [41, 44]. Comparing the peak intensities, the O1s line even exceeded the major Si2p signal, which is assigned to the elemental state of the substrate. The reasons were the relation of the atomic sensitivity factors of about $\theta_{O1s} = 0.66$ and $\theta_{Si2p} = 0.27$, respectively [61], as well as the enhanced surface sensitivity achieved by the take-off angle of 30°. Figure 3.6 also

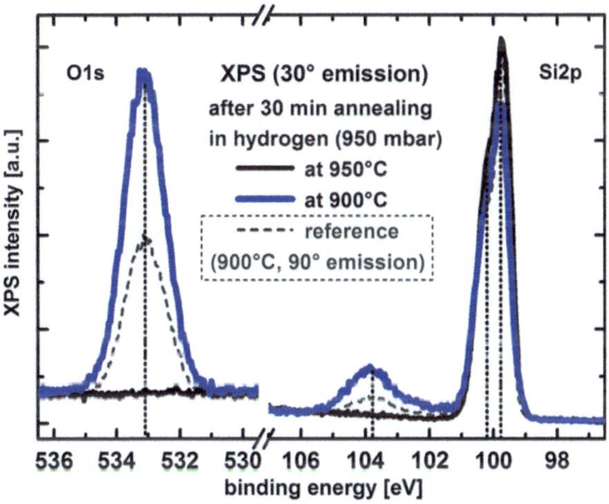

Fig. 3.6 O1s and Si2p XPS lines measured under a photoelectron take-off angle of 30° at two Si(100) samples after annealing in a MOVPE reactor under pure H_2 flow for 30 min at 950 °C (*black*, *thin*) and at 900 °C (*blue*, *thick*) surface temperature, respectively. The comparison with normalized XPS data of the sample prepared at 900 °C measured under normal photoemission (*grey*, *thin*) confirmed the expected gain of surface sensitivity (From [36])

displays normalized spectra of the 900 °C sample measured for normal photoemission (thin grey lines). In relation to the elemental Si2p peak of the Si(100) substrate, the O1s and oxidized Si2p peaks appeared with about twice the intensity in the 30° spectra (thick blue lines) compared to the 90° measurements (thin grey lines), which is consistent with a very thin SiO_2 film on the surface of the sample. In consequence, small photoelectron take-off angles provides better resolution of the XPS features associated with residual oxide layers on the Si(100) substrates after annealing in hydrogen. Comparing the Si2p spectra in Fig. 3.6 in the range of 102 eV, the signal intensity of the 900 °C sample (thick blue line) significantly exceeds that of the 950 °C one (thin black line) indicating the presence of Si in intermediate oxidation states in the residual oxide film.

3.3.2 XPS Line Shape Analysis Applied to Residual SiO_2 Layers

The line shape of the secondary Si2p peak contains detailed chemical information about the coordination of Si in the protective oxide layer or at its interface to the Si(100) substrate. Since the XPS measurements did not resolve the partial oxidation states as individual features, we conducted detailed deconvolution analysis of the Si2p spectra to reveal intermediate oxidation states of Si. Figure 3.7 shows the deconvolution results obtained for the 30° XPS measurements of a sample a sample annealed at 900 °C under hydrogen flow (same data as in Fig. 3.6). For the illustration of the Si2p photoemission spectrum in Fig. 3.7, we chose a partially logarithmic scaling of the XPS intensity to highlight the rather small contributions of oxidized Si (red and black) in the presence of the dominant signal of the elemental substrate (blue). Even in the range between the elemental and the fully oxidized Si2p peak positions, the measured XPS data (black crosses) exceeded the

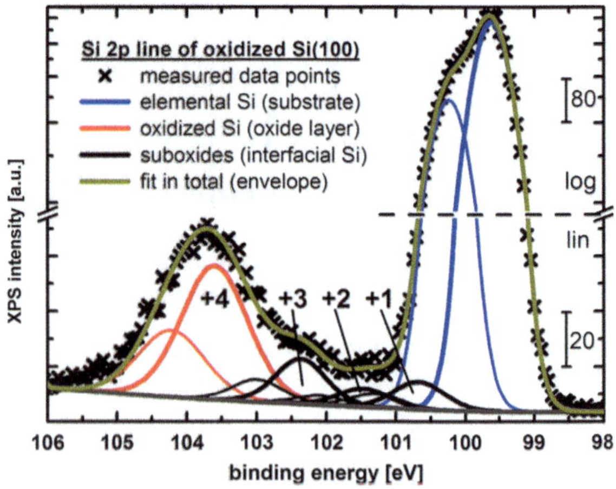

Fig. 3.7 Deconvolution analysis of the Si2p XPS signal of a Si(100) surface that is still oxidized after annealing at 900 °C in H_2. Measurement with a photoelectron take-off angle of 30° provided enhanced surface sensitivity (From [36])

expected Tougaard type background [62] (grey) significantly. The signal intensity in this intermediate binding energy range indicated some Si in partial coordination with oxygen to be considered as separate components (black) for the deconvolution analysis. Rather than ignoring the observed spin orbit coupling [43], we included it for all potential oxidation states, but fixed the parameters of the $Si2p_{1/2}$ (thinner) contributions rigorously to those of the respective $Si2p_{3/2}$ (displayed thicker) lines (Table 3.1).

The achievement of reasonable decomposition results crucially depended on deliberate handling of the different peak positions. Our deconvolution routine improved successively when entirely transferred and adjusted to several different samples until proper fits for the whole set of samples were achieved with an identical choice of the basic parameters shown in Table 3.1.

Table 3.1 specifies the parameter set utilized for the decomposition of Si2p XP spectra into Voigt profiles and exemplarily quantifies the fitting results for the experimental data as shown in Fig. 3.7. We observed a major peak at about 99.69 eV accompanied with a clearly resolved shoulder at about 100.31 eV. The splitting of the peak positions of approximately 0.62 eV and the intensity distribution agreed well with the separation of $Si2p_{3/2}$ and $Si2p_{1/2}$ due to spin orbit coupling [43]. All further peak positions attributed to stoichiometric SiO_2 as well as to intermediate suboxide states were fixed for our deconvolution model in relation to the position of elemental $Si2p_{3/2}$ (marked R). With iterative application to different samples, the peak positions of the suboxide components approached the values obtained with synchrotron radiation for UHV prepared SiO_2 layers [63]. The line width (full width at half maximum—fwhm) of the elemental Si2p components ranged close to the experimental resolution. In contrast, a reasonable fit of the SiO_2 components required significantly higher line widths due to the presence of different bonding angles [64] and phonon broadening [63].

Since we did not achieve a clear experimental resolution of the suboxide lines, their individual intensities depended to some extent on the chosen positions, line shapes and peak models. We fixed the widths of all suboxide states rather

Table 3.1 Parameters for and results of the deconvolution analysis depicted in Fig. 3.6

Si 2p deconvolution Si $2p_{3/2}$ components (oxidation state)	Binding energy (eV)	fwhm (eV)	Peak intensity ('area')	Relative intensity (of Si2p)	Relative intensity (of subox.)
Elemental (±0) [R]	99.944	0.658	685.3	0.753	–
SiO_2 (+4)	R + 3.98	1.119	147.2	0.126	–
Suboxide (+1)	R + 1.05	0.856	23.4	0.026	0.304
Suboxide (+2)	R + 1.85	0.856	15.6	0.017	0.202
Suboxide (+3)	R + 2.75	0.856	38.1	0.042	0.494
Si $2p_{1/2}$ [relative]	X + 0.62	as X	half of X	–	–

The energetic position and full width at half maximum of the utilized Voigt profiles were constrained in our model to fulfill the indicated relations between the individual components. The derived peak areas were used for the calculation of the relative intensities in relation to the total Si2p signal and to the sum of suboxide states, respectively (From [36]) arbitrarily at identical and reasonably small values [63]. In consequence, the interpretation of the relative intensities attributed to the individual suboxide states as indicated in Table 3.1 is beyond the scope of this paper. Based on our calculation of the average residual oxide thickness in Sect. 3.3, the occurrence of suboxides was attributed mainly to the $Si(100)/SiO_2$ interface. Even in the ideal case of an atomically sharp interface, the transition from a Si(100) substrate to SiO_2 film requires at least one monolayer of Si atom in partial coordination with oxygen (oxidation states other than 0 or +4) [45, 46, 65, 66]. The deconvolution results shown in Fig. 3.7 and in Table 3.1 attribute roughly one third of all oxidized silicon atoms to intermediate oxidation state, while the calculations of Sect. 3.3.3 indicate a residual oxide thickness of about 4.9 Å (Table 3.2) to the same sample under weighted consideration of the suboxide states. Given the monolayer thickness of about 1.35 Å in the Si(100) substrate and the lower concentration of Si atoms in the oxide, we draw the conclusions that

1. The observed suboxide states mainly occurred at the interface,
2. The protective oxide layers consisted of nearly stoichiometric SiO_2 and
3. Its interface to the substrate was almost abrupt.

Compared to common $SiO_2/Si(100)$ interface models [45, 46, 67], the relative fraction of the highest (+3) suboxide state (Fig. 3.7, Table 3.1) appeared higher than expected. However, the attribution of this surplus either to a specific interface model, to oxygen vacancies in the film, to initial steps of the deoxidation reaction or to the surface of the film requires dedicated experiments and high-resolution XPS as available at synchrotron facilities [45, 46, 63, 67].

3.3.3 Process Evaluation by Estimation of Oxide Film Thickness

The evaluation of the measured XPS data allowed for different estimates of the SiO_2 film thickness: according to either

Table 3.2 Oxide thickness calculation based on XPS results measured at several CVD prepared Si(100) substrates

Preparation parameters		XPS intensities [a.u.]			Relative oxygen intensity	Calculated thickness (Å)
Temperature (°C)	Time (min)	Si2p		O1s		
		Elemental	Oxidized	Total		
850	30	1,051.4	259.2	1,661.2	0.285	5.9
900	30	2,818.7	379.8	2,220.7	0.221	4.9
925	30	2,796.9	248.3	1,602.9	0.177	3.6
950	30	2,574.2	0.0	–	0.0	0.0

The variation of annealing temperatures in analogous processes with 30 min annealing time determined the critical temperature for complete removal of the protective oxide film at 950 °C (From [36])

1. The weakening of the substrate signal,
2. The relation between the O1s and Si2p intensities [61] or
3. The intensity distribution between the elemental and the oxidized Si2p peaks [68].

While the first approach presumed a homogenous layer and lacked a proper normalization, the second one required accurate knowledge of the respective atomic sensitivity factors and an adequate consideration of different photoelectron escape depths. In consequence, we relied mainly on the third approach and used the intensity of the O1s signal as independent verification.

The different chemical coordination of an element common to a thin film and the corresponding substrate can induce a chemical shift which separates two photoelectron peaks attributed to the same atomic orbital. In that case, XPS enables a straight-forward film thickness determination by analysis of the measured peak intensities [69]. The evaluation of oxidized films on silicon via the intense Si2p photoemission line [70] is a common application. For the calculation of the oxide layer thickness d_{oxy}, we determined the intensities associated with the oxidized I_{oxy} and with the elemental I_{Si} Si2p peaks from our XPS measurements (Table 3.2) and used the following equation according to Lu et al. [71]:

$$d_{oxy} = \lambda_{oxy} \sin\alpha \cdot \ln[(\beta \cdot I_{Si}) + 1] \tag{3.5}$$

The photoelectron take-off angle α and the attenuation length λ_{oxy} of the 2p photoelectrons in the oxide layer as well as the normalization factor β according to the emissions from pure bulk Si and SiO_2 represented crucial parameters. For application to the normal emission measurements ($\alpha = 90°$) at sufficiently thin SiO_2 films presented below, we applied $\beta = 0.54$ determined by measurements with our own spectrometer at clean Si(100) and a thick thermal oxide layer, respectively. This value agrees very well with theoretical predictions [68], while experimental findings are typically higher [71, 72], probably due to intensity modulations caused by photoelectron diffraction [70]. According to comparative TEM measurements in the literature [71–73], the electron mean free path in SiO_2 is about 3 nm.

Inaccurate parameter choices for thickness calculation summed up to a significant potential for systematic errors in the order of ±25 % in terms of absolute values, but the relation between differently prepared samples (Table 3.2) crucial for the evaluation of the data should be conserved. In contrast, noise in the measured XPS data also affected the peak intensities determined via the fitting routine and resulted in an uncertainty of the thickness values of individual samples smaller than ±0.3 Å. Probably, non-uniform deoxidation reaction similar to those observed in UHV [48, 50, 51] also occurred during CVD oxide removal [35], all calculated film thickness values shown in Table 3.2 only represented lateral averages over potentially inhomogeneous material distributions on the surface.

Table 3.2 shows XPS intensities for the oxidized and the elemental Si 2p as well as the O1s photoemission lines as evaluated by the systematic fitting routine described above from the data measured at several Si(100) samples from annealing experiments with variations of the applied temperature. In addition to the oxide layer thickness calculation as described above, the relative oxygen intensity in the total XPS signal (see Table 3.2) provided complementary information. Weighted by the atomic sensitivity factors $\theta_{si2p} = 0.27$ and $\theta_{O1s} = 0.66$ [61], the measured intensities represent the chemical composition of the samples. The derived oxygen content deviates from stoichiometric SiO_2 in average by approximately +20 % indication a corresponding underestimation of the residual oxide thickness. Assuming all oxygen in a thin homogenous top layer, the relative XPS intensity of the O1s line should directly correlate with the layer thickness. This rough, but independent estimation agreed very well with the calculated values based on the Si2p intensity distribution.

The experiments summarized in Table 3.2 provided a qualified evaluation of the thermal oxide removal from Si(100) in hydrogen. Annealing for 30 min at a pressure of 950 mbar and at different surface temperatures complemented the results shown in Fig. 3.6. The oxide removal rate below 950 °C was not sufficient for a complete cleaning of the substrates within the given time. The differences of the remaining oxide thicknesses (Table 3.2) between 850, 900 and 925 °C of 5.9, 4.9 and 3.6 Å, respectively, indicated increasing, but still rather small losses of oxide in this temperature range. In contrast, all samples annealed for 30 min at a surface temperature of 950 °C or above were reliably free of oxygen. Consequently, we considered 950 °C as the critical surface temperature for Si(100) substrate cleaning in the discussed environment, since even slightly lower annealing temperatures led to only partial reductions of the oxide thickness.

3.3.4 Influence of Process Gas Composition on Oxide Removal

Essentially, the previous sections described a preparation path to obtain clean Si(100) substrates by annealing in hydrogen directly within our MOVPE reactor suitable for subsequent III–V heteroepitaxy. To ensure well-defined starting conditions for the subsequent thermal oxide removal, a certain wet-chemical pretreatment strategy [15] was applied. However, the necessity to anneal the samples at about 950 °C in MOVPE environments, also observed in several previous studies [19, 27–29, 40, 52, 74], is at variances with the ability to clean Si(100) substrates in UHV by annealing at only 800 °C or even below [15]. Obviously, the distinct process gas atmosphere had a crucial impact on the Si(100) preparation.

To determine the specific preparation parameters, we changed the reactor pressure during annealing, injected small quantities of the precursor molecule silane (SiH_4), and even applied alternative process gases [36]. Table 3.3 summarizes the experimental results, evaluated in analogy to Sect. 3.3.3, in terms of the critical surface temperature required for complete removal of the protective SiO_2 layer under the specified process gas compositions.

The reduction of the process gas pressure during Si(100) substrate annealing in hydrogen from 950 to 100 mbar increased the experimental complexity of our evaluation, since the thermal coupling of the sample to the susceptor weakened

Table 3.3 Evaluation of the critical surface temperatures required for oxide removal from Si(100) substrates in dependence on the process gas compositions during the annealing process

Process gas	H_2	H_2	$H_2 + SiH_4$	N_2	Ar
Pressure (mbar)	950	100	950	950	950
T_{min}(clean) (°C)	950	900	915	–	1,000
T_{max}(oxidized) (°C)	925	870	850	970	950

For each environment, T_{min}(clean) specifies the lowest surface temperature during a 30 min annealing process for which an oxide free Si(100) surface was subsequently confirmed by XPS, while T_{max}(oxidized) refers to the highest value for which an oxide film was still detected (From [36])

simultaneously. Instead of a fairly good agreement between the heater controlled process temperature according to a thermocouple in the susceptor in our MOVPE reactor and the crucial surface temperature at nearly atmospheric pressure, an increasing discrepancy occurred at lower pressures. The in situ surface temperature measurements showed that this offset actually overcompensated the significantly higher thermal input to the susceptor required for complete oxide removal at 100 mbar: while the necessary process temperatures exceeded 1,000 °C, the actual surface temperature of the Si(100) samples was only at about 900 °C (Table 3.3). Although Si(100) preparation in MOVPE systems with closer thermal coupling might profit from the slightly lowered critical surface temperatures for oxide removal when decreasing the process gas pressure (Table 3.3), we kept the standard pressure in this study at 950 mbar to benefit from better temperature control and reduced energy consumption.

The decrease of the critical surface temperature for oxide removal with a reduction of the process pressure appears consistent with physical limitations of the deoxidation reaction due to the process gas flow [35], but also direct, chemical influence of hydrogen is conceivable. In the latter case, reactive species such as atomic, ionic or excited hydrogen as generated by the decomposition of precursor molecules should promote the progress of oxide removal. We added small quantities of SiH_4 to the process gas at a level well below growth occurred. Actually, we achieved a significant reduction of the critical surface temperature for oxide removal in the order of about 50 K (Table 3.3) when supplying SiH_4 during annealing. Besides hydrogen radicals, thermal SiH_4 decomposition certainly produces Si atoms, which also have to be considered as influence factor. The formation of volatile SiO has been established as the most relevant mechanism for Si(100) deoxidation in UHV and might be limited by the supply of elemental Si to the surface of the protective oxide layer.

3.3.5 *Impact of Alternative Process Gases on Si(100) Preparation*

For a better separation of physical and chemical influences of CVD environments on the deoxidation of Si(100), we also performed experiments using alternative process gases. Kinetic limitations of SiO desorption due to the process gas flow in a CVD ambient [31] have been proposed as the cause of the discrepancy of the required annealing temperature for oxide removal compared to UHV [15]. In that case, the type of process gas should be rather irrelevant for the deoxidation of Si(100) substrates. However, the use of nitrogen as process gas, which is an established alternative for MOVPE growth [75–78], did not clean any Si(100) substrate up to 30 min annealing time even at the maximum process temperature of our reactor, which corresponded to about 1,050 °C at the surface. Although XPS measurements indicated a slight reduction of the oxygen quantity after high temperature annealing, further experiments appeared unnecessary for Si(100) surface preparation, since we simultaneously started to detect nitrogen at the surface (Fig. 3.8).

Different explanations appeared conceivable for the presence of oxygen after Si(100) substrate annealing in N_2 (Fig. 3.8): Either the decomposition of the SiO_2 layer itself necessitated a chemically more active environment such as H_2, or the decomposition products were transported less efficiently from the surface by N_2, or the Si(100) surface simply reoxidized due to potential oxygen impurities in some part of the process environment. Apart from that, the observation of a significant N1s photoemission signal on a still oxidized Si(100) surface after high temperature annealing in N_2 (not shown here) already rejected the consideration of nitrogen as an inert process gas [75–78] during Si(100) deoxidation. Chemically, either silicon nitride formed after partial removal of the protective SiO_2 layer [47, 50] at some position at the surface, or a direct conversion of the SiO_2 film to some oxynitride phase [44, 79, 80] occurred under the conditions applied. A rough screening of N_2

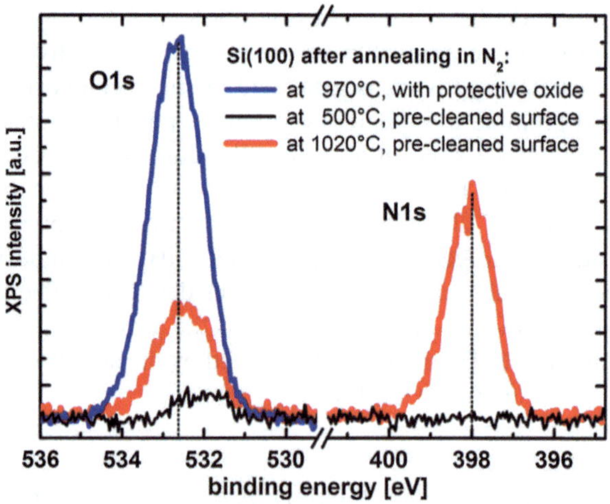

Fig. 3.8 O1s and N1s photoemission lines measured at three Si(100) substrates which were annealed in nitrogen at a pressure of 950 mbar: Two sample were previously cleaned in a hydrogen ambient and then annealed in N_2 at surface temperatures of about 500 °C (*thin black*) and 1,020 °C (*thick red*), respectively, while another one was annealed at 970 °C exclusively in N_2 process gas (*blue*) (From [36])

annealing procedures applied to already cleaned Si(100) substrates (Fig. 3.8) clarified some of these points.

After 30 min annealing at 1,000 °C in H_2, sufficient for reliable oxide removal from Si(100) (see above), the process gas was switched to N_2 for further annealing. In accordance with the regularly applied UHV transfer for surface analysis [34] which included switching to the easier pumpable nitrogen as a process gas at lower sample temperatures, prolonged annealing at a moderate value of 500 °C (Figs. 3.8 and 3.9, thin black spectra) also kept the sample essentially clean. However, increased surface temperatures of 850 °C (not shown here) and 1,020 °C (Figs. 3.8 and 3.9, thick red spectra) during N_2 annealing produced significant contaminations on the Si(100) surfaces. Since these did not only involve nitrogen but also oxygen, the N_2 used as process gas in our MOVPE reactor either itself contained oxygen impurities or at least promoted secondary oxygen sources in the environment (such as external leaks, the susceptor or even the substrates) more than H_2.

The measured O1s, N1s, and Si2p photoemission line positions and intensities (Figs. 3.8 and 3.9) enabled considerations about the actual chemical configurations at the surfaces of the samples annealed in N_2: Although the intensity ratio between the O1s and N1s varied significantly for different sample preparations (Fig. 3.8), the individual line positions remained rather constant. The nitrogen line always peaked at a binding energy of about 398.2 eV, while the oxygen position shifted almost negligibly compared to a pure SiO_2 film, at most to about 532.8 eV. Despite the consistency of both positions with the respective pure silicon compounds Si_3N_4 [44, 81, 82] and SiO_2 [41, 42], the Si2p spectra showed distinctly different behavior (Fig. 3.9).

Next to the dominant elemental peak from the substrate, the intensity distribution over the oxidized Si2p signal clearly shifted to lower binding energies. N_2 annealing after oxide removal in H_2 led to the highest N1s/O1s intensity ratios and a clear Si2p peak at about 102.4 eV (Fig. 3.9, thick red spectrum). The chemical shift of +2.9 eV compared to the elemental substrate signal exceeded the position

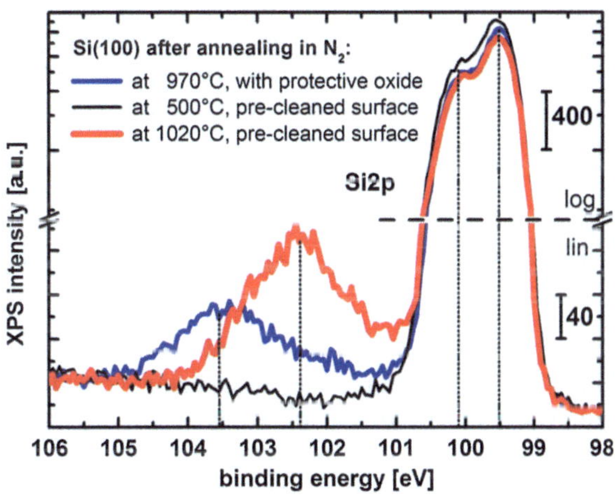

Fig. 3.9 Si2p photoemission at an partially logarithmic intensity scale measured at the three Si(100) substrates also shown in Fig. 3.8: Prior to annealing at surface temperatures of about 500 °C (*thin black*) and 1,020 °C (*thick red*) in nitrogen at a pressure of 950 mbar, both samples were cleaned in hydrogen, while the third one was directly annealed in N2 at 970 °C (*blue*) (From [36])

of 101.6 expected for pure Si_3N_4 [81, 83, 84] and rather complied with an oxynitride phase such as Si_2N_2O [84]. On the one hand, the XPS results disproved the existence of direct O–N bonds on the samples and indicated the complete coordination of both species with Si, respectively. On the other hand, the Si2p intensity distribution (Fig. 3.9, red spectrum) suggested that many silicon atoms in the surface layer were coordinated partially with oxygen and partially with nitrogen [42, 79, 80], but the results did not allow the identification of a fix oxynitride stoichiometry on our samples so far. The experiments with N_2 annealing only (not shown here) led to lower N1s/O1s intensity ratios and the Si2p line shape indicated the coexistence of SiO_2 with some oxynitride phase [42, 79].

Since Si_3N_4 formation started well below the critical temperature for oxide removal (Table 3.3), N_2 was not suitable as inert gas when used as a process gas for Si(100) preparation in a MOVPE reactor. Since we even observed oxygen contaminations after annealing of clean Si(100) surfaces in N_2, the results were not directly comparable to the preparation in purified H_2 (Table 3.3). Instead, we required a definitely inert alternative process gas to derive conclusions about the influence of the process gas on thermal oxide removal. The installation of argon at the carrier gas supply enable the investigation of Si(100) deoxidation in a noble gas. Actually, annealing for 30 min at a pressure of 950 mbar at a surface temperature of about 1,000 °C removed the protective SiO_2 layer (Table 3.3). However, the annealing temperature of 950 °C failed to clean Si(100) substrates in Ar, while it was reliably sufficient for surface preparation in H_2 under identical conditions (Table 3.3).

The critical surface temperature for oxide removal in the CVD environment determined from our experiments appear essentially consistent with the deoxidation mechanism in UHV [48, 50, 51] and the lower preparation temperature required there [15] under the assumption of a physical restriction of SiO desorption due to the process gas flow [31]. In hydrogen, the critical temperature for oxide removal decreases with pressure. In argon at identical process pressure, also high temperatures were required, which even slightly exceeded the values for hydrogen. Besides that, also the results obtained by SiH_4 injection to the process gas might support the existence of an additional deoxidation mechanism assisted by the presence of hydrogen, but do not provide clear evidence. We currently plan detailed experiments to characterize the interaction of a hydrogen-based CVD ambient with Si(100) surfaces directly using in situ optical and post preparation infrared spectroscopy.

3.4 Variation of Essential Substrate Parameters

In the previous sections, we carried out an analysis of the thermal oxide removal from Si(100) in our MOVPE reactor considering the essential preparation parameters independently of the type of substrate. A certain wet-chemical pretreatment strategy [15] was employed in order to ensure well-defined starting conditions, but we also conducted comparative experiments to investigate the dependency of the substrate cleaning procedure on the type of surface termination prior to CVD processing. Due to the possibility of diffusion of residual oxygen contained in the substrates to the surface during annealing, we also considered the type of crystal manufacturing technique as a relevant parameter for the ability to remove the oxide.

3.4.1 Comparison Between Wet-Chemical and Native Oxide Layers

The control over undesired impurities when applying different strategies of wet-chemical Si(100) substrate pretreatment strategies essentially depends on the prevention of subsequent recontaminations from the ambient atmosphere. The lack of reproducibility of H-terminated Si(100) surfaces produced by a final HF-dip justified the protection of cleaned substrates by an intentional, wet-chemical formation of a well-defined oxide layer suitable for removal in UHV at relatively low temperature [15]. In contrast, several studies reporting on III–V growth on Si(100) and utilizing the same pretreatment for MOVPE preparation of the substrates [19, 27–29, 40, 52, 74] employed significantly higher temperatures of around 950 °C to clean the samples, in agreement with our observations (see previous sections). Several comparative oxide removal experiments using both natively and chemically oxidized samples (Table 3.4) addressed the question, whether results of the Si(100) preparation in a CVD environment depended on the type of oxide layer.

Our experiments did not indicate any substantial difference of the oxide removal when using native or chemical oxide layers on Si(100) substrates. At a pressure of 950 mbar, annealing in H_2 always led to identical results (Table 3.4): For each of many experiments, XPS detected either the absence of oxygen on both substrate types or quite similar residual oxide layer thicknesses. Even with different preparation environments, only very few discrepancies in the residual oxide content were observed: The difference between the two samples processed in parallel (with identical preparation parameters) in H_2 at a pressure of 100 mbar most probably resulted from coincidence with slightly worse thermal coupling of the susceptor to the natively oxidized sample. At this low pressure, significant offsets between the susceptor temperature and the actual value at the surface were

Table 3.4 Comparison of the residual oxide thicknesses d_{ox} calculated according to XPS measurement according to Eq. 3.5 between Si(100) samples with native and wet-chemically prepared SiO_2 layers various CVD processes (From [36])

Annealing process	None	T low	Standard	P low	SiH_4 injection
H_2 pressure (mbar)	–	950	950	100	950
Surface temperature (°C)	–	850	950	925	850
Chemical oxide: d_{ox} (Å)	8.3	5.7	0.0	0.0	4.0
Native oxide: d_{ox} (Å)	8.2	6.6	0.0	7.4	8.9

commonly detected (Sect. 3.3.4), while only the surface temperature of one of the samples could be monitored by an optical in situ probe. Besides reducing the critical temperature for oxide removal, admixture of small SiH_4 quantities to the standard H_2 process gas also led to slightly different residual oxide films after annealing at 850 °C. However, here the homogeneity of the native oxide films was not checked before performing the experiments.

3.4.2 Comparison of Differently Manufactured Si(100) Substrates

Although none of the experiments resulted in a strong indication of an influence of the oxide type on the oxide removal, we carried out the 'Shiraki' procedure [15] as our standard for further experiments to ensure a well-defined contamination control and proper comparability between results from all samples. In future experiments, XPS investigations of structural differences between different SiO_2 films or their interfaces with the substrate require a higher resolution of the individual sub-oxide peaks within the Si 2p photoemission line [45, 46, 63] as is only achievable with synchrotron radiation as light source. Our experiments so far show that, eventually, the omission of wet-chemical pre-treatment might be an economical option for future production processes, if supply with natively oxidized substrates without other impurities can be guaranteed.

The need for successful and complete thermal oxide removal for the preparation of Si(100) surfaces in CVD environments did not only involve obtaining a thin, well-defined SiO_2 top layer, but also included oxygen impurities in the bulk of the substrate [85, 86] which potentially diffused during annealing. Crystal growth processes utilized in Si wafer manufacture of course incorporate certain quantities of light impurities: Common 'Czochralski' (CZ) silicon typically contains an oxygen concentration of about 10^{18} cm^{-3}, while the elaborate 'float zone' (FZ) process allows a significant reduction. Considering potential costs for industrial production, we chose CZ Si(100) as standard substrates for our investigations and obtained FZ Si(100) manufactured from identical basic material for comparison. Oxide removal from both types of substrate conducted in H_2 at 950 mbar for various annealing temperatures did not result in any obvious differences. Secondary ion mass spectroscopy (SIMS) provided a quantitative measure for the oxygen concentration contained in the investigated substrates (Fig. 3.10).

Apart from sputtering artifacts from the oxidized surfaces, quantitative SIMS depth profiles showed the constant intrinsic oxygen concentration levels of untreated CZ and FZ substrates (Fig. 3.10, black lines) of about 9.2×10^{17} cm^{-3} and about 5×10^{16} cm^{-3}, respectively.

Moderate H_2 annealing at 850 °C, which was insufficient to remove the protective oxide layer, induced significant changes in the oxygen distribution in the bulk of a CZ Si(100) sample (Fig. 3.10, grey line). Within a distance of some hundreds of nanometers from the surface, a considerable reduction of the impurity

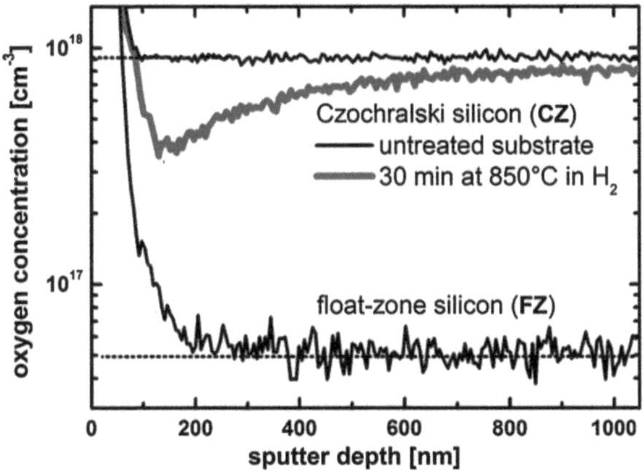

Fig. 3.10 Quantitative SIMS measurements of the oxygen concentration in different Si(100) substrates plotted over sputtering depth. Higher concentrations at the onset of the spectra are due to SiO_2 layers at the surface and the well-known effect of recoil implantation. Higher measurement depths confirm the different intrinsic oxygen concentration levels associated to CZ (*top*) and FZ (*bottom*) manufacturing of Si. The curve recorded after thermal annealing in H_2 shows a thermally activated trend for the diffusion of oxygen from the bulk to the surface of the samples

level (up to a factor of three) was observed, which gradually returned to the bulk level measured in untreated silicon substrates. The result corresponded to a total loss of oxygen in the bulk near the surface, which can only be explained by oxygen diffusion to the Si(100) surface [85–87]. However, the effective amount of oxygen which diffused to the surface remained negligible in comparison with that already contained in the protective oxide layer at the surface. The material depths affected by oxygen diffusion increased significantly for higher annealing temperatures, but the formation of an additional SiO_2 monolayer would have required the oxygen impurity content of a several μm thick layer at a concentration observed in the bulk of CZ silicon. Although our results excluded a decisive impact of oxygen diffusion on the thermal oxide removal from the surface, the annealing caused significant changes of the intrinsic impurity level in the material [85–87].

3.5 Atomic Surface Structure After Thermal Oxide Removal

The choice of crucial parameters during Si(100) substrate manufacturing already sets the stage for the atomic surface structure after thermal removal of the protective SiO_2 layer. On the one hand, the chosen degree of wafer misorientation determines the principal step density of the cleaned Si(100) surface. On the other hand, the off-cut direction with respect to the symmetry of the bulk lattice simultaneously determines the potential interplay between the step edges and the surface reconstruction mechanism [9]. Due to undesirable anti-phase boundary initiation at single (or odd-numbered) layer steps [2, 3], the systematic formation of double layer steps [12, 88] on the substrate surface prior to III–V on Si(100) hetero-epitaxy represents a crucial challenge for the CVD preparation of Si(100) [27]. Since the type of Si(100) step structures observed in UHV critically depends on the principal step density, we chose 0.1°, 2° and 6° wafer misorientations toward the desirable [011] direction [10] to obtain a representative investigation [37].

3.5.1 Surface Reconstruction and Domain Distribution

In principle, dimer formation on Si(100) surfaces [7] induces a (2 × 1) symmetry. Since single-layer surface steps separate terraces with perpendicular dimer orientation [7, 8] resulting in a characteristic two-domain (2 × 1)/(1 × 2) surface reconstruction. The systematic formation of double layer steps theoretically explained by Chadi [9] and experimentally observed for vicinal surfaces with high step density [10] is equivalent to a single-domain reconstruction. For all Si(100) substrate types which were prepared by thermal oxide removal in hydrogen, LEED measurements showed clear diffraction patterns (Fig. 3.11) indicating clean surfaces with a well-ordered atomic structure.

Figure 3.11 displays a comparison of the LEED patterns typical of the investigated Si(100) wafer misorientations directly after thermal oxide removal in the MOVPE reactor, each measured with the identical incident electron energy of 100 eV. The diffraction pattern of the only slightly off-cut substrate (Fig. 3.11a) is essentially identical to the distinctive two-domain (2 × 1)/(1 × 2) surface reconstruction well-known from UHV prepared Si(100) [7, 89]. Equally intense half-order diffraction spots (see arrows in the LEED images) in both surface

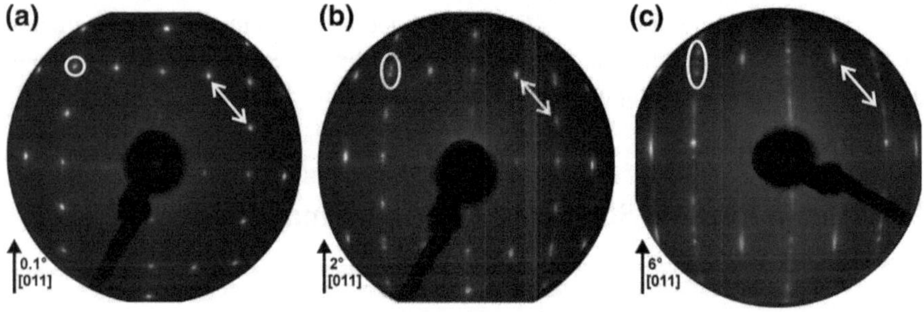

Fig. 3.11 Typical 100 eV LEED patterns observed at Si(100) substrates, misoriented by 0.1° (**a**), 2° (**b**) and 6° (**c**) in the [011] direction, after CVD oxide removal: In (**a**) and (**b**), typical half-order diffraction spots in both directions confirmed a two-domain (2 × 1)/(1 × 2) surface reconstruction. However, some fine structure due to the periodicity of the step edges also occurred for higher off-cut angles and interferes with evaluation of the domain content, especially in (**c**) (From [37])

directions indicate the presence of both surface reconstruction domains without any obvious preference for either. Principally, terraces of perpendicular dimer orientation can only be separated by step configurations with odd-numbered atomic step height [9], but these steps also initiate undesirable anti-phase disorder during the heteroepitaxial growth of typical compound semiconductors on such substrates [2, 3, 27].

On the vicinal Si(100) substrates with 2° (Fig. 3.11b) and 6° (Fig. 3.11c) misorientation, we observed some additional fine structure of the diffraction spots in the off-cut direction (encircled in Fig. 3.11). Here, the LEED spots in [011] direction split due to the periodicity given by the distance of the step edges on the surface [90] and thereby reflected the average reciprocal terrace width. The clear observation of the splitting confirmed a high regularity of the actual step structures. Judging from the LEED patterns, the utilized CVD procedure resulted in smooth Si(100) surfaces without any indication for step bunching or faceting. As in the case of the almost exactly oriented Si(100), the LEED patterns of CVD-cleaned vicinal substrates with 2° off-cut (Fig. 3.11b) always showed half-order diffraction spots in two crystallographic directions, which correspond to the existence of undesirable single layer steps. For 6° off-cut (Fig. 3.11c), the inequivalent intensity distribution in the LEED patterns might indicate some imbalance between the surface reconstruction domains, but the splitting of the diffraction spots in only one direction, of course, interfered with a thorough observation of the spot intensities. For a clearer impression of the half order spots in the [011] direction of the 6° off-cut sample of Figs. 3.11c, 3.12 shows LEED images measured at the same surface, but with different primary energies.

In the LEED images shown in Fig. 3.12, the regions, where the half order spots in the [011] direction are expected, are highlighted by white circles. At these positions, the spot intensities are dependent on

1. The presence of residual (1 × 2) surface reconstruction domains,
2. The spot splitting according to 6° off-cut, and
3. The incident electron beam energy.

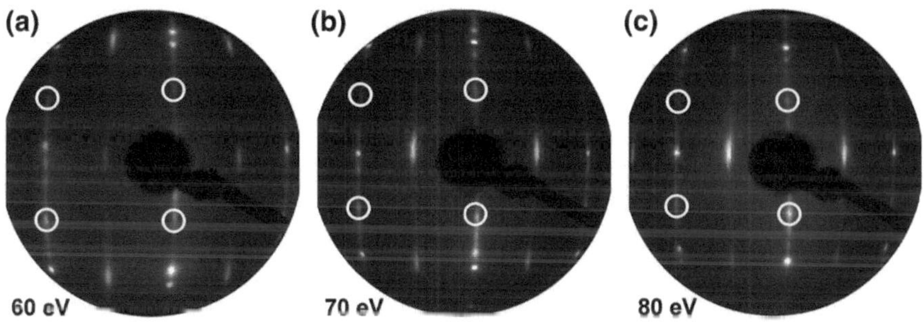

Fig. 3.12 Typical LEED patterns observed at the Si(100) substrate of Fig. 3.11c with 6° misorientation in the [011] direction with incident beam energies of 60 eV (**a**), 70 eV (**b**), and 80 eV (**c**): The expected positions for half-order diffraction spots in the [011] direction are indicated by white circles (From [37])

In fact, we always observed half order contribution to the LEED pattern in the [011] direction with varying intensity. For example, pronounced (1 × 2) spots can be seen at the lower left-hand position of the 60 eV (Fig. 3.12a) as well as at the lower centre position of the 70 eV (Fig. 3.12b) and 80 eV (Fig. 3.12c) images. Hence, energy-dependent LEED observations at 6° off-cut Si(100) samples after oxide removal in our MOVPE reactor provided evidence of the presence of residual (1 × 2) surface reconstruction domains in the [011] direction.

Although LEED was not suitable to quantify the distribution of surface reconstruction domains on vicinal substrates, the presence of a residual, second domain on our CVD-prepared Si(100) even for larger degrees of misorientation did not agree completely with standard UHV results for clean surfaces [25]. Producing single-domain Si(100) surfaces with double-layer steps via oxide removal in UHV by simple thermal treatment is well-known and well-established for sufficiently large misorientations [12, 88]. A possible source of discrepancies between the different preparation procedures in the different environments, UHV versus CVD, originated from the physical and chemical interaction of the surface of the sample with or without the process gas. Presumably, hydrogen bonds at the surface changed the energetics of dimer formation, step configurations and surface diffusion. Since both the technological aim [2] as well as the general potential [9] for single-domain Si(100) are based on the atomic configuration of the step structure, further considerations required a microscopic analysis of the surfaces.

3.5.2 Atomic Structure and Surface Morphology

On the microscopic level, scanning probe techniques provided insight into the actual Si(100) surface structures. While subsequent ex situ AFM enabled rather rough inspections of the morphology on a μm scale, UHV-based STM, accessible by a direct transfer from the MOVPE reactor to UHV [34], allowed to analyze the crucial step configurations of the Si(100) surfaces in much more detail down to atomic resolution. The choice of 0.1°, 2° and 6° misorientations towards the [011] direction implied basic conditions with very different step densities. Many studies [7, 10, 12, 25] have discussed the consequences for the UHV preparation of Si(100) and served as detailed references for our results obtained after thermal oxide removal in a CVD environment.

3.5.2.1 Si(100) with Slight Misorientation (0.1° Towards [011])

Assuming only single-layer steps, the slight off-cut of only 0.1° determined an equivalent average terrace width of about 80 nm for these Si(100) substrates in good agreement with our observations. Figure 3.13 shows a typical area of a substrate deoxidized by our standard H_2 annealing procedure (Sect. 3.2). Mono-layer steps prevailed in the microscopic surface morphology and separated

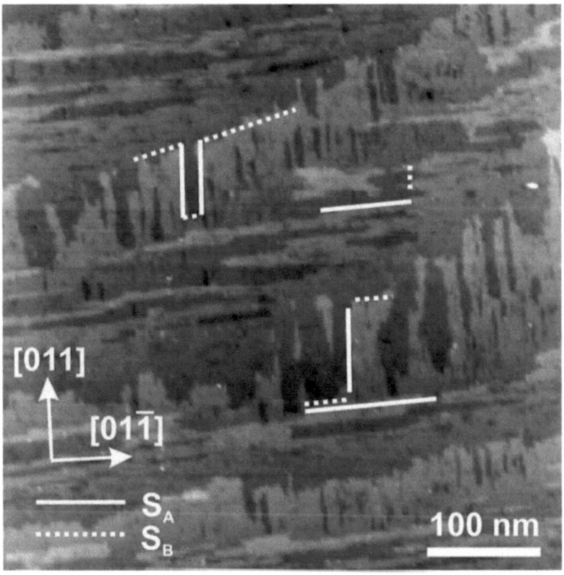

Fig. 3.13 Empty state image (V_{Sample} = +1.2 V, I_t = 400 pA) of a thermally deoxidized Si(100) surface on a substrate with 0.1° misorientation toward [011]: Frayed monolayer steps dominate the surface structure, the alternating directionality allows straightforward differentiation between S_A and S_B type steps (From [37])

domains attributed to perpendicular dimer orientations. Although Fig. 3.13 did not provide atomic resolution, differences in the arrangement of adjacent terraces are obvious: All terrace edges are frayed, but the appearance changes direction between adjacent terrace edges. Elongated fringes occur on every other terrace perpendicular (in [011] direction) to the general step direction, while they are parallel (in $[0\bar{1}1]$ direction) on the terraces in-between. In analogy to theoretical results [9, 91, 92], different step formation energies depending on the dimer orientations at the step edge probably determined the driving force for the microscopic arrangement.

Due to the symmetry of the diamond crystal structure, the binding orientation alternates for each monolayer and the Si(100) surface reconstruction mechanism induces a change of the dimer orientation between adjacent terraces unless double-layer steps are formed. In the notation established by Chadi [9], single layer steps with dimers on the upper terrace oriented either perpendicular (S_A) or parallel (S_B) to the step edge occurred. Calculations of the step formation energy [9] favor the formation of S_A type steps over the S_B type steps on clean Si(100) in UHV. Actually, the situation depicted in Fig. 3.13 agreed well with the preferential elongation of one of these step types (solid lines) due to frequent interruption of the other one (dashed lines). Compared to the basic step configurations on (2 × 1)/(1 × 2) reconstructed Si(100) surfaces described by Chadi [9], differences in the step structure occurred in our studies due to interactions with the chemical environment (process gas) and a potential H-termination of the surface [21]. High resolution STM imaging (Fig. 3.14) eventually proved that the elongated step type seen in Fig. 3.13 featured a dimer orientation perpendicular to the step edge on its upper terrace. Consequently, the

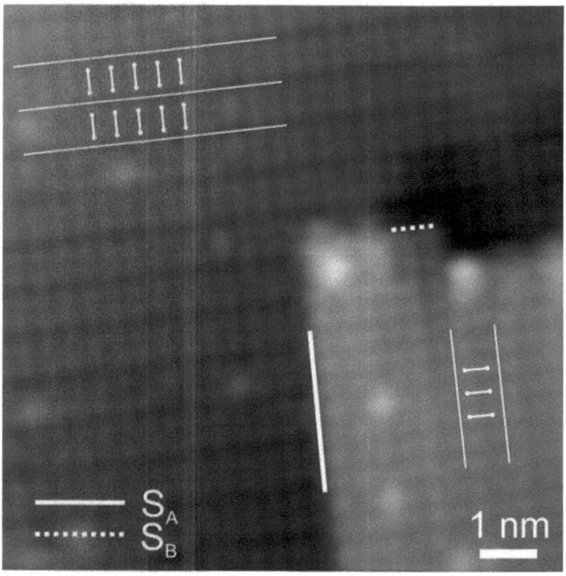

Fig. 3.14 Atomically resolved empty state image ($V_{Sample} = +1.3$ V, $I_t = 50$ pA) of thermally deoxidized Si(100) with 0.1° misorientation in [011] direction: At the monolayer step, the direction of the Si surface dimers changes. According to the definition of step types, the assignment switches between S_A and S_B where kinks occur, as illustrated in the image (From [37])

energetic preference for S_A type steps represented a direct analogy to clean Si(100) without H-termination in UHV as described by Chadi [9].

In detail, Fig. 3.14 shows the microscopic vicinity around an S_A type step with atomic resolution. The resolution of individual Si atoms (indicated as dots) and variations of their distance in bonding direction enabled the reliable identification of Si dimers [23, 93]. Perpendicular to the respective bonding direction, the atomic distances remained constant, which led to the impression of characteristic Si dimer rows (indicated as lines). At the single atomic step edges, the orientation of the dimers (and dimer rows) changed to the perpendicular directions. Corresponding LEED measurements (Sect. 3.1) displayed averaged information of surfaces of similar samples (Fig. 3.11a) and showed the superposition of surface reconstruction domains with (2 × 1) and (1 × 2) symmetry induced by the respective dimer orientations on the terraces. The classification of the step in Fig. 3.14 as S_A type resulted from the single atomic height (S) and perpendicular orientation of the dimers relative to the step edge on the upper terrace (type A) [9].

Since step edges observed in our STM studies on Si(100) surfaces with 0.1° off-cut substrates neither ran straight nor parallel, contrary to the expectation based on the manufacturer's specifications, we also performed systematic inspections of the surface morphology on larger length scales with subsequent ex situ AFM (Fig. 3.15). Typical images obtained after thermal removal of the protective oxide layer (Fig. 3.5a) showed that the step edges were roughly oriented in the $[0\bar{1}1]$ direction (from left to right in Fig. 3.15a) in agreement with the projection of the misorientation in the [011] direction according to the specification. However, the step structures appeared irregular and slightly curved. The ridges with a typical distance of 1 to 1.5 μm extending in the [011] direction were a result of several

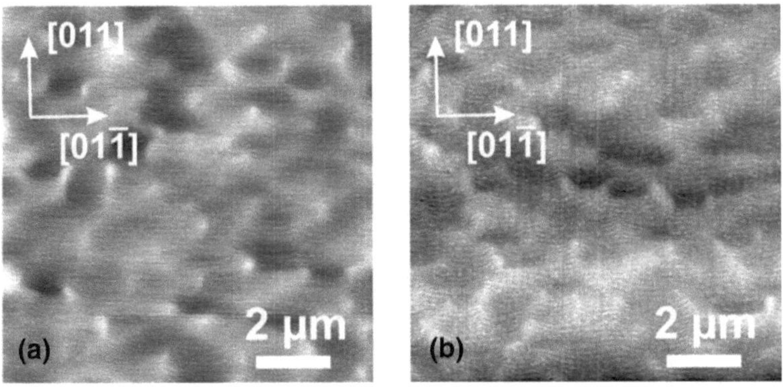

Fig. 3.15 AFM images of deoxidized Si(100) surfaces: Both substrates were 0.1° misoriented towards [011], but prior to thermal oxide removal in H_2 they were covered either with a wet-chemical protective oxide layer (**a**) or with the native oxide of the wafer (**b**) (From [37])

steps merging. In order to compare different preparation routes, we carried out AFM measurements on samples prepared by the identical CVD procedure from the same substrate, but without wet-chemical pre treatment (Fig. 3.15b). Figure 3.15a, b show qualitatively similar surface morphologies after thermal removal of the different oxide films. Therefore, the wet-chemical etching procedure could be ruled out as the origin of the ridges, which were rather attributed to polishing artifacts from wafer manufacturing or perhaps to non-uniform deoxidation processes [31, 94].

The microscopic surface structures observed on 0.1° off-cut Si(100) substrates directly after CVD surface preparation appeared consistent with UHV results in the literature. However, Ref. [27] reports on preferential double step formation on quite similar substrates in a CVD environment, but the sample preparation involves a specific buffer growth and annealing procedure. The step structure of our deoxidized Si(100) as depicted in Fig. 3.13 already indicates a significant difference in step formation energy depending on the step type and, therefore, a potential origin for an imbalance of surface reconstruction domains induced by growth. However, the preparation of single-domain Si(100) surface structures on slightly misoriented substrates in a MOVPE reactor reported in Ref. [27] still represents a distinct difference to our CVD and standard UHV results [12, 24, 25, 88]. According to the theoretical calculations by Chadi [9], on clean Si(100) the formation energies of double-layer steps with Si dimers parallel (D_B) to the step edges are more favorable than D_A steps (Si dimers perpendicular to step edges) or combinations of S_A and S_B steps. STM results from UHV prepared samples confirmed the suggested double-layer step formation mechanism, if the step density exceeded a value corresponding to a substrate misorientation of around 2.5° [12, 88].

3.5.2.2 Si(100) with Intermediate Misorientation (2° Towards [011])

Our LEED results obtained after deoxidation of vicinal Si(100) substrates excluded the preparation of purely single-domain surfaces: we found two almost equally weighted domains on 2° (Fig. 3.11b) and the presence of a residual minority domain on 6° (Figs. 3.11c and 3.12) substrates. A thorough evaluation required deeper insight into the actual step structures. In fact, the thermal oxide removal from 2° Si(100) samples led to a complex surface structure involving roughening on a μm scale (Fig. 3.16a). With higher spatial resolution, we observed varying, but highly regular step structures (Fig. 3.16b), which were not always directly correlated to the general misorientation of the Si(100) substrate in terms of step density as well as step orientation.

Figure 3.16a exhibits an array of flat terraces separated by areas with non-uniform morphology. Some of the terraces show clearly visible step structures with typical step widths of a few tens of nm. Due to its limited resolution, the STM image of Fig. 3.16a did not allow a precise determination of the step heights, but their straight appearance did not agree with the frayed structure of S_B type steps

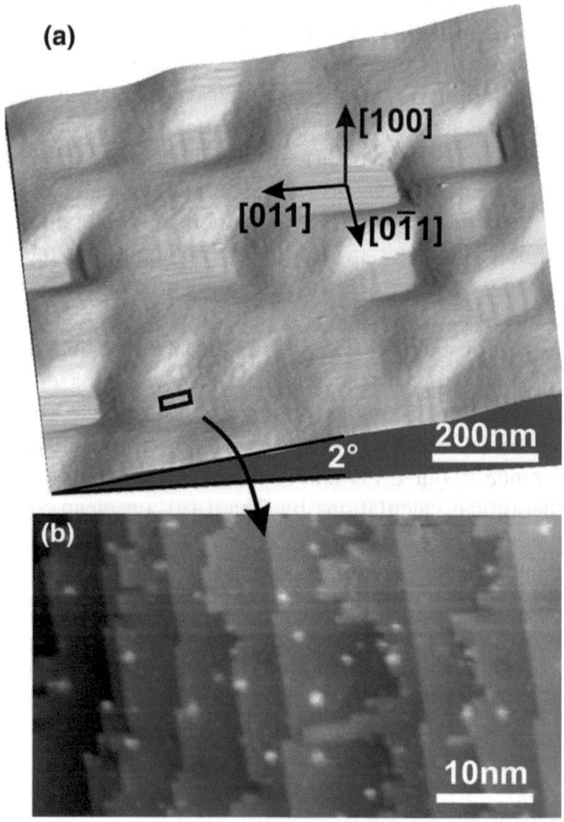

Fig. 3.16 Empty state images of thermally deoxidized Si(100) with 2° misorientation in [011] direction: The large scale image (**a**) (V_{Sample} = +1.2 V, I_t = 60 pA) shows a typical array of terraces separated by rougher regions in-between. In the high-resolution image (**b**) (V_{Sample} = +1.5 V, I_t = 50 pA), the step structure of one of the rougher regions of part (**a**) is visible in more detail (From [37])

shown in Fig. 3.13. Hence, Fig. 3.16a represents an indirect indication for partial double step formation, since single stepped areas necessarily involve alternating S_A and (frayed) S_B steps. Several of the terraces are entirely flat without any structure, which, in contrast to the other terraces, suggests that they are exactly (100) oriented. In areas separating the terraces in Fig. 3.16a, the local misorientations differ significantly in direction and magnitude, which of course affects the local step density and orientation as well as the appearance of the possible step types.

We recorded several high resolution STM images at different positions between the terraces visible in Fig. 3.16a, which clearly resolved the local step structure, as exemplarily shown in Fig. 3.16b. Here, the local gradient is in the direction of the specified substrate misorientation, but the magnitude of about 1.1° is lower than expected. Two types of steps, appearing alternately in the image, can clearly be distinguished: every other step appears straight, almost without kinks, while the steps in-between look rather frayed. The step heights are consistent with monolayer steps and, as in the case of the 0.1° misoriented samples (Fig. 3.13), we can identify the straight steps with the S_A type, while the frayed ones correspond to the S_B type. The terraces separated by the two different step types are of similar size, leading to a similar distribution of dimers oriented in [011] and [0-11] direction, in good agreement with the LEED observation of a two-domain surface (Fig. 3.11b).

3.5.2.3 Si(100) with High Misorientation (6° Towards [011])

Figure 3.17 shows STM images of a sample with 6° off-orientation. In the main image, steps with irregular spacing are visible, which do not correspond to mono- or double-layer steps but to step bunches of irregular height and distance. Due to this irregular local distribution of steps, the surface is not smooth on a lateral scale of around 1 μm, with height variations in the order of 3 nm. The inset shows a small region of the surface with higher resolution, where dimers are visible, but, due to a non-ideal tip, atomic resolution comparable to the images shown before was not achieved for this sample after thermal oxide removal in hydrogen. The image shows predominantly double layer steps, but these include many defects, particularly missing dimers and missing dimer rows. The dimer rows are oriented perpendicular to the step edges, and therefore the steps are of the D_B type. In order to clarify the atomic structure of the 6° Si(100) surface we also characterized similar samples after growth of a thin homoepitaxial buffer layer (Fig. 3.18).

The STM image in Fig. 3.18 shows an irregular step structure of a 6° Si(100) surface with atomic resolution. It is characterized mainly by double-layer steps, but single-layer and triple-layer steps occur in several places. In Fig. 3.18, single- and triple-layer steps are indicated by broken white lines, with the remainder of the step edges belonging to double-layer steps. Single- and triple-layer steps necessarily delimit terraces of the residual (1 × 2) surface reconstruction domains, where dimer rows run parallel to the step edge, as opposed to the majority (2 × 1) domain, where they run perpendicular. Hence, Fig. 3.18 does not show a single-domain Si(100

Fig. 3.17 Empty state images of thermally deoxidized Si(100) with 6° misorientation in [011] direction: The large scale image (V_{Sample} = +1.3 V, I_t = 70 pA) shows step structures with irregular spacing, while the inset (V_{Sample} = +0.9 V, I_t = 520 pA) depicts a region with predominantly double layer steps in high-resolution where the Si dimers are clearly visible (From [37])

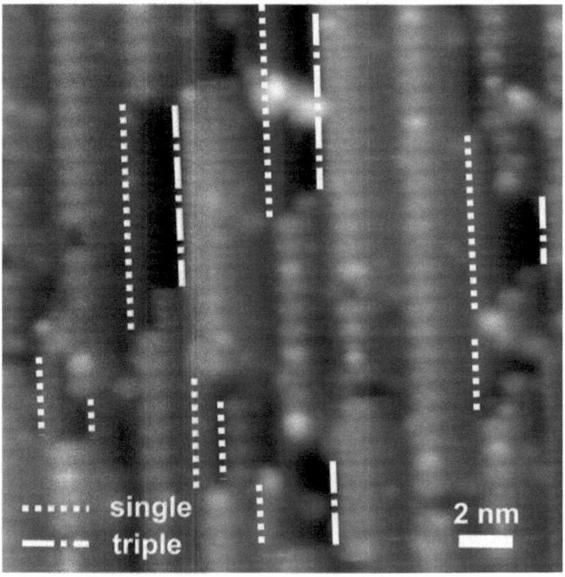

Fig. 3.18 Empty state images (V_{Sample} = +1.3 V, I_t = 100 pA) of a Si(100) substrate with 6° misorientation in [011] direction after growth of a thin homoepitaxial buffer layer: The irregular step structure is characterized mainly by double-layer steps. Individual single-layer and triple-layer steps are indicated by *dotted* and *dot-dashed white lines*, respectively (From [37])

surface, but suggests the presence of a minority domain covering a surface fraction of about 10–20 %. Besides buffer growth, this 6° off-cut sample was prepared identically to that of Fig. 3.17. Lower resolution STM images of the samples of Figs. 3.17 and 3.18 showed very similar surface structures, and the superstructures in their LEED images agree very well. Therefore, the image in Fig. 3.18 is likely to provide a good indication of the atomic surface structure of 6° Si(100) prepared in the CVD environment. In summary, the surface exhibits a tendency towards double layer steps, but is far from ideal, especially in comparison with the UHV-prepared clean Si(100) surface at large misorientations [12, 88].

3.6 The Role of Hydrogen During CVD Surface Preparation

Advanced control over Si(100) surface structures represents a crucial requisite for the heteroepitaxial integration of compound semiconductors with silicon [1], which is desirable for an increasing number of technological applications [3]. While detailed knowledge about clean Si(100) and its preparation in ultra high vacuum (UHV) results from extensive research activities [8], optoelectronic III–V devices are commonly manufactured by metal–organic vapor phase epitaxy (MOVPE). Only few studies [21, 22] consider the interaction of Si(100) surfaces with hydrogen-based process ambients typical for chemical vapor deposition (CVD) processing, even though it represents the common standard for high-performance device production.

Simultaneously, the potential influence of hydrogen on the atomic surface structure of Si(100) substrates [25] is of major interest: during heteroepitaxial integration of polar compound semiconductors on non-polar Si(100) surfaces, single-layer steps of the substrate initiate undesirable anti-phase boundaries [2], while the formation of double-layer steps, predicted by surface energy calculations [9], suppresses anti-phase disorder [1]. According to experimental observations in UHV, the preference for double-layer steps prevails at least for vicinal Si(100) with sufficiently high step densities [10, 12], whereas exposure to atomic hydrogen represses the occurrence of double-layer steps [25]. Although chemical vapor environments typically involve hydrogen as carrier gas and as precursor fragment, Kunert et al. [27] recently reported a VPE preparation route for Si(100) resulting in a predominance of double-layer steps on specific substrates with only slight off-orientation in [011] direction.

Hence, there is no consensus on the interaction of hydrogen from CVD environments with Si(100), neither regarding the impact on oxide removal and surface preparation, nor regarding the influence on the atomic surface structure and the step configuration. In this section, we use several surface sensitive instruments to substantiate the presence of hydrogen on CVD-prepared Si(100) substrates [38]: Fourier-transform infrared spectroscopy (FTIR) showed coupled Si–H bonds and verified a nearly single-domain surface according to the dominance of one dimer orientation; scanning tunneling microscopy (STM), resolving individual Si-dimers, proved the complete hydrogen saturation of the dangling bonds by tip induced hydrogen desorption; and in situ reflectance anisotropy spectroscopy (RAS) showed the absence of the hydrogen termination during the thermal annealing steps associated with the CVD preparation of Si(100) surfaces.

3.6.1 Verification of Si–H Bonds on CVD-Prepared Si(100)

The variation of the process gas between pure H_2, diluted SiH_4 and pure Ar during thermal annealing of Si(100) substrates indicated an active role of hydrogen in the removal of the protective oxide layer (Sect. 3.3). Eventually, we performed Fourier-transform infrared spectroscopy (FTIR) to obtain direct evidence for an interaction of the H_2 process atmosphere with the substrates. After wet-chemical treatment [15], thermal annealing at 1,000 °C under flow of purified hydrogen at nearly atmospheric pressure (950 mbars) in our MOVPE reactor removed the protective SiO_2 layer. Following growth of a homoepitaxial buffer layer by exposure to the precursor silane, a dedicated sample transfer involving a mobile UHV chamber [34] enabled contamination-free access to several remote surface analysis systems, as confirmed by X-ray photoelectron spectroscopy [35, 95]. A commercial FTIR spectrometer (Bruker IFS 66v/S) was used to conduct hydrogen-sensitive ATR measurements in a dedicated UHV chamber designed to fit into the sample compartment of the spectrometer. For further characterization of the interaction of the process gas with the Si(100) surfaces during CVD preparation, we investigated similarly prepared samples by STM (Specs 150 Aarhus STM) and by optical in situ spectroscopy (LayTec EpiRAS 200).

If the sample environment contained reactive hydrogen species, they would react with the cleaned Si(100) surface and form rather stable Si–H bonds observable as characteristic absorption lines by FTIR spectroscopy. An attenuated total reflection (ATR) configuration enabled surface sensitive FTIR measurement [96], but required a special sample geometry (ATR crystal) realized in a low doped material. For this study, an ATR crystal manufactured from 2° misoriented vicinal Si(100) substrate directly served as the sample [97].

Figure 3.19 shows FTIR spectra measured at the wet-chemically cleaned sample [15] before (grey) and after (black) annealing at 950 °C in pure H_2 at 950 mbar. The analysis presented in Fig. 3.19 based on single channel spectra and the application of a linear background as reference for the relevant range [97], since no adequate reference sample was available. Before the annealing procedure, a protective oxide layer [15] covered the Si(100) ATR substrate and the spectra (Fig. 3.19, red line) contained no indications for hydrogen bonds in the crystal or at its surface. After thermal oxide removal in H_2, additional absorption lines showed up (Fig. 3.19, black line) in the range characteristic for Si–H vibrational modes [21, 55]. Apparently, some reaction of the process gas with the sample formed a considerable amount of Si–H bonds during annealing. The long optical path of the IR signal due to multiple internal reflections induced a high sensitivity for relatively rare chemical bonds compatible with surface species or with very low concentrations in the bulk. The required temperature for oxide removal in H_2 could have enabled hydrogen diffusion into the substrate and reactions with impurities or defects. However, the two absorption lines observed at about 2,099 and 2,088 cm^{-1} (Fig. 3.19) complied

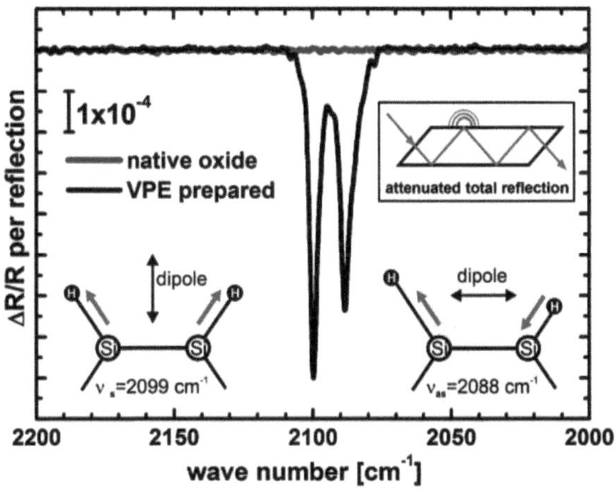

Fig. 3.19 Surface sensitive FTIR spectra measured in an attenuated total reflection (*ATR*) configuration at a Si(100) sample before (*grey*) and after (*black*) thermal oxide removal in H_2. Clear features emerged in the region around 2,100 cm^{-1} and can be assigned to the symmetric and asymmetric stretch modes of coupled Si-monohydrides (After [110])

very well with the symmetric (2,099 cm^{-1}) and the antisymmetric (2,087 cm^{-1}) stretch modes of the coupled Si–H monohydrides as observed at Si(100) surfaces after atomic hydrogen supply in UHV [55].

3.6.2 Polarization Dependence of the ATR Spectra

The assignment of the Si–H bonds to the prepared Si(100) surface assumes its termination by H–Si–Si–H dimers. However, pure, hydrogen free Si(100) also reconstructs by the formation of an analogous dimer configurations, only terminated by dangling bonds instead of hydrogen. Both the observation of (2 × 1)/(1 × 2) symmetry by LEED (Sect. 3.5.1) as well as the resolution of individual Si-dimers by STM (Sect. 3.5.2) on Si(100) substrates similar to the ATR crystal used for FTIR could not differentiate atomic surface configurations with and without hydrogen, and, in consequence, could not confirm either of the cases. The characteristic coupling of the coupled monohydride stretch modes to polarized IR radiation dependent on the orientation of the H–Si–Si–H dimers [26, 98] provided direct evidence for the assignment of all Si–H bonds to the Si(100) surface by the FTIR measurements shown in Fig. 3.20.

The ATR data measured at our VPE-prepared Si(100) samples showed two major absorption peaks at positions assigned to the characteristic stretch modes of coupled Si–H monohydrides [98] in agreement with previous CVD studics [21, 22]. The adsorption of atomic hydrogen at the residual dangling bonds of a typical (2 × 1)/(1 × 2) reconstructed Si(100) surface leads to H–Si–Si–H dimers vibrating

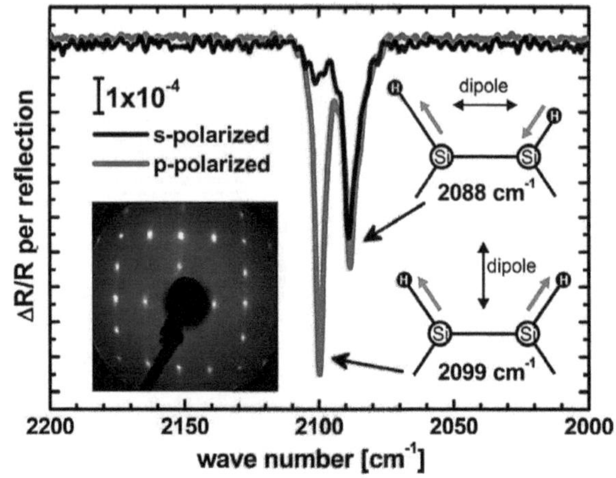

Fig. 3.20 ATR-spectra of VPE-prepared Si(100) employing s-polarized and p-polarized radiation and associated LEED pattern (*inset*) of the same sample: Absorption at 2,099 and 2,088 cm^{-1} is assigned to the symmetric and antisymmetric stretch modes of the Si–H monohydride, respectively. In agreement with the $(2 \times 1)/(1 \times 2)$ symmetry observed by LEED, almost equal intensity of the antisymmetric stretch mode in both s- and p-polarized spectra indicates a two-domain Si(100) surface (After [110])

in symmetric and antisymmetric modes and absorbing at about 2,099 and 2,088 cm^{-1}, respectively. The visibility of the lines in ATR spectra measured with polarized IR radiation (Fig. 3.20) depends on the orientation of the induced electric dipole moments [98].

Figure 3.20 shows two ATR spectra in the spectral region of the characteristic Si–H absorption bands measured at the same CVD-prepared Si(100) sample, but with different polarization. As sketched in the inset, the dipoles associated with the symmetric stretch mode are perpendicular to the surface and therefore only couple to p-polarized radiation, while the dipoles of the antisymmetric stretch mode are oriented within the surface plane and potentially couple to both polarizations. Since almost equal absorption was observed with either polarization direction [55], the quantity of monohydride terminated Si-dimers oriented in the plane of incidence was comparable to those with perpendicular orientation. In agreement with the associated $(2 \times 1)/(1 \times 2)$ LEED pattern (Fig. 3.20, inset), the ATR result indicated the preparation of a two-domain Si(100) surface reconstruction.

3.6.3 Verification of Nearly Single-Domain Si(100)

The variation of our VPE surface preparation procedure to slower cooling in the presence of molecular hydrogen eventually enabled us to prepare almost single-domain Si(100) surfaces (Sect. 3.7). In contrast to the polarized ATR measurements

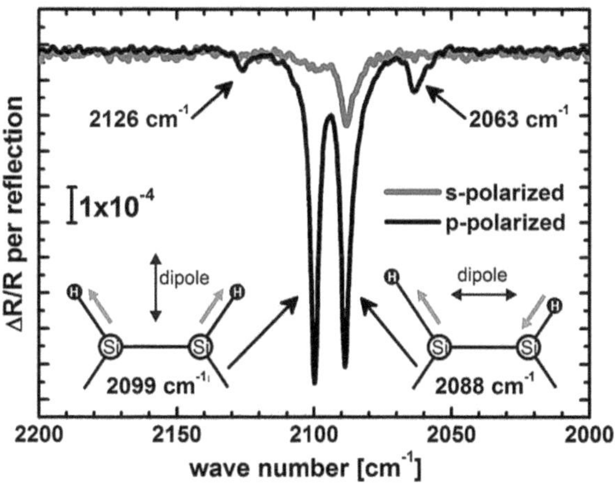

Fig. 3.21 ATR-spectra of nearly single-domain Si(100) prepared by VPE employing s-polarized and p-polarized radiation: Absorption at 2,099 and 2,088 cm^{-1} is assigned to the symmetric and antisymmetric stretch modes of the Si–H monohydride, respectively. The different intensity of the antisymmetric stretch mode in both s- and p-polarized spectra indicates a nearly single-domain Si(100) surface. Minor absorption lines at 2,099 and 2,088 cm^{-1} are probably due to hydrogen bonds at double-layer step edges (After [110])

at CVD-prepared Si(100) of Refs. [21, 22], our data [38] depicted in Fig. 3.21 shows intense absorption at 2,088 cm^{-1} for p-polarization, but only a weak line for s-polarization, instead of very similar intensities that were observed in these studies. The effect is due to the distribution of dimer orientations in correlation with the relative extent of the associated surface reconstruction domains.

The results shown in Fig. 3.21 indicate the preparation of a nearly single-domain Si(100) surface reconstruction with most H–Si–Si–H dimers oriented parallel to the plane of incidence and only few oriented in the perpendicular direction. Since their antisymmetric stretch modes couple almost equally to radiation in the respective polarization directions [55], the intensity ratio between both absorption lines directly reflects the domain distribution. Based on the data presented in Fig. 3.21, we estimate the residual content of the minority surface reconstruction domain on our sample as less than 20 %. Besides the coupled monohydride vibrations, Fig. 3.21 also shows two minor absorption lines at about 2,063 and 2,126 cm^{-1}, respectively. Since we did not observe these on explicitly two-domain Si(100) samples (not shown here), an association to characteristic Si–H configurations located at the edges of double-layer steps [99] appears very likely.

3.6.4 Confirmation of a Complete Monohydride Termination

The visibility of Si-dimers in ATR spectra depends on their saturation with hydrogen. Reference [21] shows a distinct dependence of the peak intensities associated with coupled monohydrides on the hydrogen partial pressure during sample preparation without observing any additional Si–H bands. Accordingly, saturated and unsaturated dimers may coexist on a Si(100) surface. Since ATR spectroscopy does not allow a reliable quantification of hydrogen bonds, these measurements cannot provide evidence for a complete monohydride termination of the entire Si(100) surface. Hence, we addressed the question of the hydrogen coverage by separate STM experiments (Fig. 3.22) conducted on an identically prepared Si(100) substrate, which featured a lower off-orientation of 0.1° in order to obtain larger terraces.

Figure 3.22a shows an STM image with atomic resolution, where Si-dimers and their orientation are easy to identify. As STM images do not provide direct information on hydrogen coverage, we attempted to desorb hydrogen using sample bias voltages in excess of +3 V, as described by Shen et al. [57]. Figure 3.22 displays a comparison of two images of the same area recorded before (a) and after (b) a scan at +3.4 V. Several bright features appeared; in particular, some bright double lobes are visible in Fig. 3.22b. These are typical of pairs of dangling bonds on a silicon surface, where pairs of hydrogen atoms have been desorbed from an otherwise monohydride-terminated surface [100]. We tentatively attribute the smaller, individual bright features in the image to single dangling bonds. Our experiment provides strong evidence for a complete saturation with hydrogen bonds as a consequence of the utilized preparation procedure, since no dangling bonds were observed at the VPE-prepared Si(100) surface before the STM desorption experiment, while they appeared as very distinct features at a few

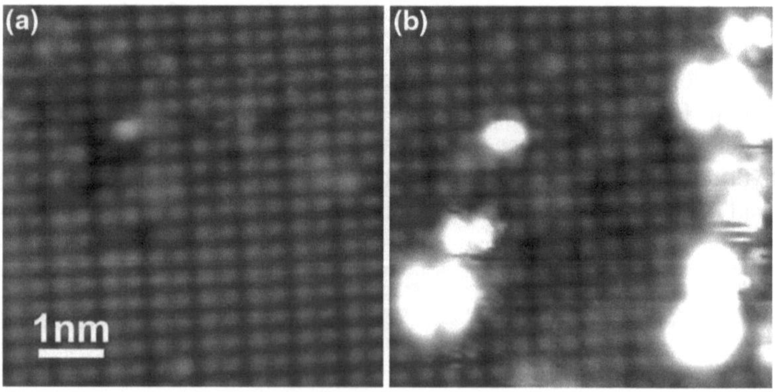

Fig. 3.22 Empty states STM images (sample bias +1.2 V) of MOVPE-prepared Si(100) before (**a**) and after (**b**) tip-induced hydrogen desorption: bright double-lobed features outshining the regular surface structure are indicative for pairs of dangling bonds (From [38])

places thereafter. However, the actual reaction parameters leading to the monohydride termination still remained unclear. On the one hand, atomic hydrogen as used by Chabal et al. [26] in UHV occurred as an essential byproduct of SiH_4 decomposition during growth of the homoepitaxial buffer layer. On the other hand, Aoyama et al. [21] showed the replacement of a wet-chemical deuterium termination by a monohydride termination during moderate hydrogen annealing ($p \leq 32$ mbar; $T \leq 800\ °C$) and consequently discuss the adsorption of molecular hydrogen at unsaturated Si-dimers.

3.6.5 In Situ Observation of Si(100) Hydrogenation in the Process Gas

We performed optical in situ spectroscopy to obtain direct experimental evidence for the hydrogen coverage of our Si(100) samples during the VPE-preparation (Fig. 3.23) using substrates with a higher off-orientation of 6° towards [011]. According to well-established RA signatures of Si(100) prepared in UHV [56, 101, 102], we expected clearly distinguishable spectra of clean and monohydride terminated vicinal surfaces suitable for in situ monitoring.

Figure 3.23 shows characteristic in situ RAS spectra of the Si(100) surface at crucial steps of the VPE process. During thermal oxide removal by annealing at 950 °C in hydrogen, we observed the development of an intense, wide minimum between 2.2 eV and 3.8 eV (Fig. 3.23, thick grey line) consistent with the spectrum recorded after Si buffer growth. However, cooling of the sample in the process ambient was accompanied by a major change in signal shape towards the ambient temperature spectrum (Fig. 3.23, black line) with a narrower peak structure and a steep rise of the signal around the E1 inter band transition at about 3.4 eV. Similar results have been reported in the literature [56, 102] after atomic hydrogen exposure at room temperature in UHV-based environments:

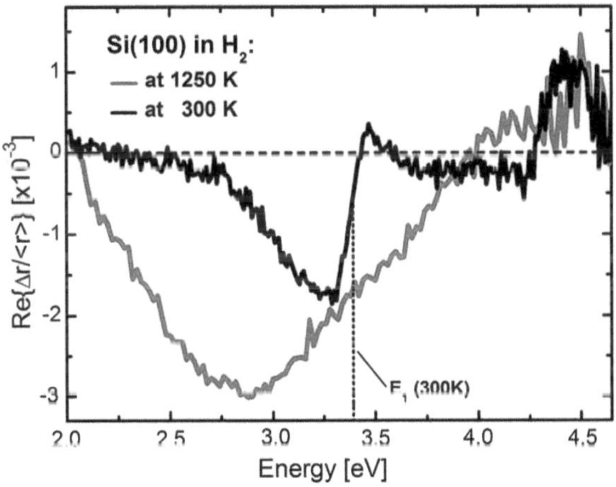

Fig. 3.23 In situ RAS spectra of VPE-prepared Si(100), 6° off-oriented in [011]: directly after deoxidation at 1,250 K in hydrogen, the surface clearly shows the well-known, wide feature of a pure substrate, while the in situ signal after cooling to 300 K in the same ambient confirms a monohydride termination of the surface (From [38])

1. Our RAS observation at elevated annealing temperature (Fig. 3.23, thick grey line) agrees very well with spectra assigned to clean vicinal Si(100) surfaces, in particular if slight thermal shifts of the peak position and coupling of thermal radiation into the spectrometer are taken into account.
2. In analogy to atomic hydrogen adsorption, cool-down of the Si(100) surface under H_2 flow led to a final RA spectrum (Fig. 3.23, black line) associated with monohydride termination directly confirmed by correlation with our ATR results (Fig. 3.21).

The RAS data shown in Fig. 3.23 demonstrate the direct optical in situ observation of the hydrogenation during Si(100) preparation in an MOVPE reactor. At the annealing temperature, the hydrogen desorption rate most likely outbalances the process of dissociative adsorption of molecular hydrogen at the surface. Therefore, we observed an effectively hydrogen free Si(100) surface, which potentially represents a characteristic equilibrium of the dynamic processes in the MOVPE environment—in contrast to the Si(100) surface prepared by atomic hydrogen adsorption at ambient temperature in UHV. Subsequently, during cooling in the process environment, the surfaces most probably pass a transitional temperature regime, where desorption is already suppressed, while adsorption still occurs—and the complete monohydride termination confirmed at ambient temperature is formed. Due to our in situ RAS observations, the transition from the clean to the monohydride spectrum started at about 800 °C and deserves detailed analysis in future experiments.

3.7 Unconventional Double-Layer Step Formation in a CVD Ambient

We investigated the preparation of double-layer steps on 0.1°, 2° and 6° misoriented Si(100). The samples with 0.1° misorientation were prepared by a similar process to that reported by Németh et al. [103] and Kunert et al. [27] They obtained predominantly double layer steps when the off-orientation was in [011] direction, but intermediate monolayer terraces appeared when the off-orientation was different from any of the ⟨011⟩ directions. As they used AFM for their investigations, the limited resolution did not allow them to differentiate between D_A and D_B type steps. Also, they could not image the step edge with high resolution, so that they were unable to verify the complete absence of an intermediate terrace in the case of off-orientation in [011] direction. We used STM to investigate this type of surface more closely [39], enabling the detection of both the terrace type as well as the microscopic structure at the step edge.

3.7.1 Si(100) with Slight Misorientation (0.1° Towards [011])

Figure 3.24 shows the step structures of a region on a 0.1° Si(100) surface similar to that shown in Ref. [27, 103]. In Fig. 3.24a, regularly spaced terraces are visible, separated by double-layer steps, where, as in the case of the AFM images of Ref. [27, 103], no intermediate step edges are resolved. However, the image clearly shows vacancy islands on the terraces, which form trenches extending roughly parallel to the step edges. Figure 3.24b shows a typical area of a terrace with larger magnification, allowing analysis of step edges and vacancy islands in more detail. The terraces are delimited by smooth edges, and measurement of the vertical separation of the terraces confirmed a separation corresponding to two atomic layers, so it appears that this image provides evidence of double-layer steps.

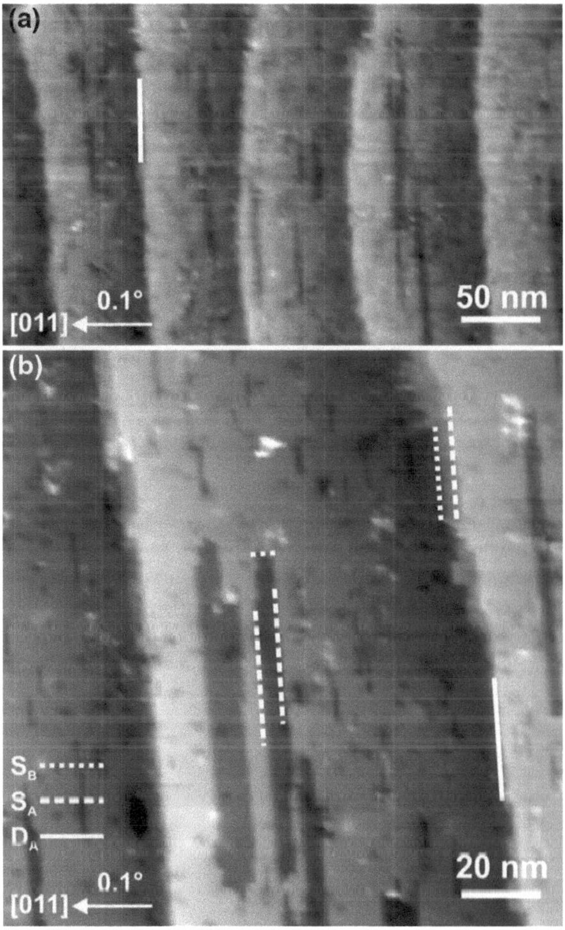

Fig. 3.24 **a** Empty state image of the step structure of a 0.1° misoriented Si(100) surface, which appears to be characterized by double-layer steps of the D_A type (V_{sample} = 780 mV, I_t = 700 pA). **b** Magnified section of a double layer step, which in most parts consists of an S_A step and an S_B step, where the intermediate terrace is reduced to a *narrow strip*, and in some location is missing altogether, resulting in true D_A type steps. On the terraces, single layer vacancy islands exist, which are elongated parallel to the terrace edges, leading to long S_A type and short S_B type vacancy edges (V_{sample} = 1,240 mV, I_t = 290 pA)

Close inspection of the step edge, however, shows the presence of a narrow intermediate terrace. The edges of the main terraces are mostly smooth, indicative of an A-type step, while those of the intermediate terrace are frayed, as is expected for a B-type step. Hence the step structure, in most locations, does not consist of a truly double-layer step, but a combination of an S_A and an S_B step, separated by a very narrow intermediate terrace, as indicated in one position in Fig. 3.24a. In some stretches of the step edges in this figure, no intermediate terraces are visible, but the evidence of a truly D_A type step is not conclusive due to the limited resolution of the image, i.e. an intermediate terrace consisting of a few dimers cannot be ruled out. On the terraces, again, many defects can be observed. While there are many smaller vacancies and a few islands in some locations, the most striking features on the terraces are, again, large vacancies which are elongated parallel to the terraces. They are delimited by single layer steps, with long step edges of the S_A and short step edges of the S_B type, in accordance with the above-mentioned (Sect. 3.5.2.1) preference for the S_A type steps. These microscopic measurements are consistent with the AFM measurements of Yanase et al. [104] who reported the observation of S_A type steps, with saw teeth-like S_B steps in-between, which were clearly visible at a sample of 0.1° misorientation where the steps edges slightly deviated from the $\langle 011 \rangle$ direction, but could not be resolved in an image at 0.2° misorientation where the steps were parallel to the $\langle 011 \rangle$ direction.

In order to investigate the step structure in more detail, we imaged one of the terrace edges shown in Fig. 3.24a with atomic resolution, so that dimer rows are clearly visible (Fig. 3.25). The image shows two terraces, separated in most parts by an intermediate ledge. On the terraces, the dimer rows are parallel to the terrace edge,

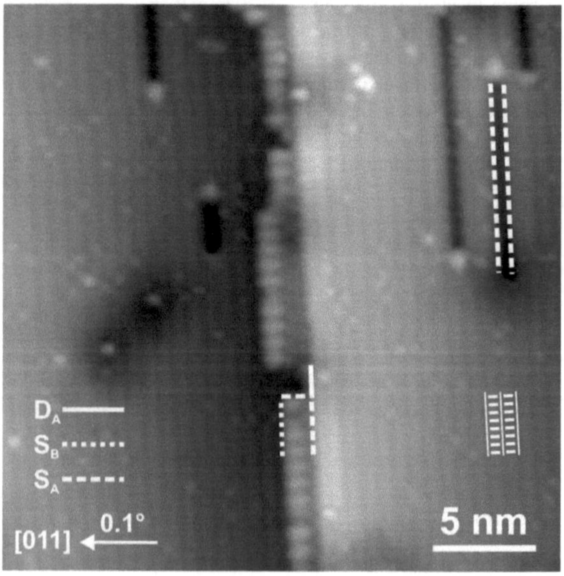

Fig. 3.25 Atomic resolution empty state image of pseudo-D_A type step on a 0.1° misoriented Si(100) surface. In most parts, the step consists of an S_A step and an S_B step, where the intermediate terrace is only a few dimers wide. On the terraces, vacancy islands consisting of missing dimer rows are visible ($V_{sample} = 760$ mV, $I_t = 840$ pA)

as is expected for an A-type edge. The intermediate ledge is only a few dimers wide, but the exact microscopic structure of the termination of the dimer rows forming the ledge is not obvious from the STM image. The image of Fig. 3.25 confirms the observation based on Fig. 3.24, that in most parts the step edge of the 0.1° surface is not truly of the double layer D_A type, but consists of a combination of S_A and S_B steps, where the intervening terrace is reduced to an extremely narrow ledge. Nevertheless, the surface exhibits a strong preference for the domain where dimer rows run parallel to the step edges, so that subsequent III–V growth should result in a small minority domain, where APDs in the epilayer have a high probability to disappear due to APB self-annihilation over a short distance.

In the atomic resolution image of Fig. 3.25, a few defects are visible. Of these, the most pronounced features are missing dimer row vacancies, terminated by brighter features, most likely impurity atoms, which pin the end of the vacancy. This type of defect may be compared to the vacancy islands in Fig. 3.24. In Fig. 3.24b, narrow vacancies of similar size can also be seen. The large vacancy islands in Fig. 3.24b have the same orientation and could originate from the same process.

Possibly, the larger vacancies are the result of the diffusion and coalescence of mobile smaller islands which in contrast to those visible in Fig. 3.25 have not been immobilized by defects. In the left-hand terrace two small protrusions are visible, each seemingly surrounded by an indentation. They are probably due to charged impurity atoms at the surface of the sample. The indentations are not a topographic feature, but rather due to a reduced density of states in the vicinity of the impurity atoms.

3.7.2 *Si(100) with Intermediate Misorientation (2° Towards [011])*

FTIR investigations have shown that our surfaces are terminated by Si–H monohydrides. According to literature, a tendency toward double layer steps of the D_A type is unexpected for the hydrogenated Si(100) surface. Theoretical analysis of step formation energies [92, 105] find a preference for single layer steps on this surface, in contrast to the clean surface where double layer steps of the D_B type are energetically favorable. Most experimental studies are in agreement with these theoretical results: Laracuente and Whitman [25] find a preference for single layer steps at the monohydride surface for misorientations up to 7°. Others mostly report on preparation of smooth, single-layer stepped surfaces, resulting form annealing in hydrogen at temperatures on the order of 1,000 °C [106, 107].

Only a few studies mention direct observation of double layer steps at the Si(100) surface prepared in the presence of hydrogen: Kitahara and Ueda [108] report on a tendency for a retreat of S_A like steps on 2° misoriented samples. Nemeth et al. [103] provided evidence of a tendency towards double layer steps at lower misorientation, where the type of step was not specified due to the limitation of AFM imaging. Their data is similar to that of Yanase et al. [104], which was also obtained on a sample annealed in hydrogen atmosphere. These results are completely consistent with our measurements at the 0.1° misoriented hydrogenated surface: this surface consists of large terraces with dimer rows parallel to the step edge and very narrow terraces in-between, where the dimer rows run perpendicular to the step edge, and which in some places disappear altogether, resulting locally in true D_A steps. In Ref. [38] we reported on FTIR measurements on a Si(100) sample with 2° misorientation in [011] direction. In the polarized spectra, an imbalance in the domain ratio was detected, where the majority domain consisted of dimer rows parallel to step edges (the orientation of the steps was evidenced by LEED measurement). The proportion of the minority domain on this sample was on the order of 20 %, based on the relative intensity of the absorption lines due to the antisymmetric stretch modes of the dimers oriented parallel and perpendicular to the step edges, respectively.

In contrast to previous samples, the MOVPE process used in the preparation of this sample contained a cooling phase from 1,000 to 500 °C extending over 1 h as a final step in the process. We repeated this procedure on several samples and obtained similar results, which were in sharp contrast to those obtained by the process used in the preparation of 2° samples reported by us in Ref. [37]. The latter process resulted in a two-domain surface characterized in most parts by regions with single-layer steps, interspersed by an array of flat terraces.

In order to confirm the highly desirable domain imbalance found by FTIR in the former process, we carried out STM measurements on this type of surface. The resulting STM image in Fig. 3.26 shows a step structure of terraces with rather straight edges. The difference in height of these terraces corresponds to two atomic

Fig. 3.26 High resolution empty state image of step structure on a 2° misoriented Si(100) surface consisting largely of D_A type steps. In some parts, residuals of the intermediate terrace, consisting of dimer rows of varying length, extend from the A-type step onto the terraces (V_{sample} = 1,240 mV, I_t = 190 pA)

layers. Close inspection of the terraces shows dimer rows parallel to the step edges, about 0.4 nm wide. In many locations at the step edges, short dimer rows of variable length extend perpendicular from the edge on the lower terraces. We obtained similar images of somewhat lower quality at several locations on the sample within the scanning range of our STM (2 × 2 μm^2). The fact that the dimer rows on the terraces extend parallel to the step edges implies that the step structure is dominated by D_A steps. While in some locations along the step edges, the short intermediate ledges consisting of short dimer rows exist, there are quite a few places where a sharp two-layer step edge prevails. This result is of great technological relevance because the almost perfect dominance of one domain is ideal for subsequent APD-free III–V growth.

3.7.3 Si(100) with High Misorientation (6° Towards [011])

Based on the above results for 2° misorientation we attempted to transfer the same preparation procedure to 6° misoriented Si(100) samples, again using STM to study the resulting surface structure. Figure 3.27 shows an atomic resolution image of this surface. It displays a rich step structure containing all of the major step types which occur on Si(100). Continuous type A steps run from top to bottom of the image and, but, on most parts of the resulting terraces, dimer rows extend from the type A step edge, perpendicular to the edge, onto the terraces. Here, the step edge is of the S_A type. The perpendicular dimer rows, in some cases, terminate at some position near the center of the terrace, resulting in S_B steps. In a

Fig. 3.27 Atomic resolution empty state image of 6° misoriented Si(100) surface. The step structure consists of a mixture of S_A, S_B, D_A and D_B type steps, with no clear preference for one of the step types. (V_{sample} = 1,240 mV, I_t = 190 pA)

few places the dimer rows reach the edge of the terrace underneath, so that a D_B step is obtained. D_A steps occur where the perpendicular dimer rows are missing altogether. The image is too small to provide a reliable estimate of the domain ratio, but it is clear that a strong dominance of one domain can be ruled out. The preparation of double layer steps of the D_A type contrasts with previous theoretical [92, 105] and experimental [25, 106] work on the hydrogenated Si(100) surface. On the clean Si(100) surface, double layer steps of both types can be prepared by electromigration. However, this technique is not applicable to the hydrogenated surface.

D_A steps have been prepared in UHV on clean Si(100) by Bedrossian and Klitsner [109] as well as Swartzentruber et al. [12] using low-energy Xe ion bombardment. The method is based on the preferential removal of the surface atoms from the S_B type step due to the generation of vacancies on the terraces which easily diffuse towards and annihilate at the S_B step. This is based on evidence that vacancy islands diffuse much more rapidly in the direction of the dimer rows than in the perpendicular direction. We propose that in our case the mechanism for formation of D_A steps is also based on the removal of material. In the temperature range between 1,000 and 700 °C, the surface is not passivated by the hydrogen atmosphere and the surface silicon atoms are still mobile. If in fact silicon is removed from the surface, the resulting surface structure depends on the relation between vacancy generation and diffusion. These two processes both slow down with decreasing temperature, but their relative rates may change, so that their action on the resulting surface will depend on the rate of cooling, since it determines the onset of hydrogen passivation. If diffusion parallel to the dimer rows is much faster than vacancy generation, annihilation at the S_B step edge will dominate and the associated terrace will be removed very effectively. We conjecture that this is the case in the temperature regime close to the onset of hydrogen passivation in the case of the 2° sample.

References

1. S.F. Fang, K. Adomi, S. Iyer, H. Morkoc, H. Zabel, C. Choi, N. Otsuka, J. Appl. Phys. **68**, R31–R58 (1990)
2. H. Kroemer, J. Cryst. Growth **81**, 193–204 (1987)
3. H. Döscher, T. Hannappel, B. Kunert, A. Beyer, K. Volz, W. Stolz Appl, Phys. Lett. **93**, 172110 (2008)
4. W.T. Masselink, R. Fischer, J. Klem, T. Henderson, P. Pearah, H. Morkoc Appl, Phys. Lett. **45**, 457–459 (1984)
5. W.T. Masselink, T. Henderson, J. Klem, R. Fischer, P. Pearah, H. Morkoc, M. Hafich, P.D. Wang, G.Y. Robinson Appl, Phys. Lett. **45**, 1309–1311 (1984)
6. R. Fischer, N. Chand, W. Kopp, H. Morkoc, L.P. Erickson, R. Youngman, Appl. Phys. Lett. **47**, 397–399 (1985)
7. J.E. Griffith, G.P. Kochanski, Crit. Rev. Solid State Mat. Sci. **16**, 255–289 (1990)
8. Z.Y. Zhang, F. Wu, M.G. Lagally Annu, Rev. Mater. Sci. **27**, 525–553 (1997)
9. D.J. Chadi, Phys. Rev. Lett. **59**, 1691–1694 (1987)

10. R. Kaplan, Surf. Sci. **93**, 145–158 (1980)
11. T.W. Poon, S. Yip, P.S. Ho, F.F. Abraham, Phys. Rev. Lett. **65**, 2161–2164 (1990)
12. B.S. Swartzentruber, Y.W. Mo, M.B. Webb, M.G. Lagally, J. Vac. Sci. Technol. A-Vac. Surf. Films **7**, 2901–2905 (1989)
13. H.N. Waltenburg, J.T. Yates, Chem. Rev. **95**, 1589–1673 (1995)
14. M. Morita, T. Ohmi, E. Hasegawa, M. Kawakami, M. Ohwada, J. Appl. Phys. **68**, 1272–1281 (1990)
15. A. Ishizaka, Y. Shiraki, J. Electrochem. Soc. **133**, 666–671 (1986)
16. D.A. Neumayer, J.G. Ekerdt, Chem. Mat. **8**, 9–25 (1996)
17. H. Döscher, T. Hannappel, J. Appl. Phys. **107**, 123523 (2010)
18. M. Akiyama, Y. Kawarada, K. Kaminishi, Jpn. J. Appl. Phys. Part 2—Lett. **23**, L843–L845 (1984)
19. M.K. Lee, D.S. Wuu, H.H. Tung, Appl. Phys. Lett. **50**, 1725–1726 (1987)
20. L. Samuelson, P. Omling, H.G. Grimmeiss, J. Cryst. Growth **68**, 340–344 (1984)
21. T. Aoyama, K. Goto, T. Yamazaki, T. Ito, J. Vac. Sci. Technol. A-Vac. Surf. Films **14**, 2909–2915 (1996)
22. H. Bender, S. Verhaverbeke, M. Caymax, O. Vatel, M.M. Heyns, J. Appl. Phys. **75**, 1207–1209 (1994)
23. J.J. Boland, J. Vac. Sci. Technol. A-Vac. Surf. Films **10**, 2458–2464 (1992)
24. A. Laracuente, L.J. Whitman, Surf. Sci. **476**, L247–L253 (2001)
25. A.R. Laracuente, L.J. Whitman, Surf. Sci. **545**, 70–84 (2003)
26. Y.J. Chabal, K. Raghavachari, Phys. Rev. Lett. **53**, 282–285 (1984)
27. B. Kunert, I. Nemeth, S. Reinhard, K. Volz, W. Stolz, Thin Solid Films **517**, 140–143 (2008)
28. C.A. Tran, R.A. Masut, P. Cova, J.L. Brebner, R. Leonelli, J. Cryst. Growth **121**, 365–372 (1992)
29. H.S. Kim, Y. Kim, M.S. Kim, S.K. Min, J. Cryst. Growth **92**, 507–512 (1988)
30. M. Mayusumi, M. Imai, S. Nakahara, K. Inoue, J. Takahashi, T. Ohmi Solid State Phenom. **65–66**, 229–232 (1999)
31. T. Yamazaki, N. Miyata, T. Aoyama, T. Ito, J. Electrochem. Soc. **139**, 1175–1180 (1992)
32. M. Grundmann, A. Krost, D. Bimberg, J. Vac. Sci. Technol. B **9**, 2158–2166 (1991)
33. T. Hannappel, W.E. McMahon, J. M. Olson J. Cryst. Growth **272**, 24–29 (2004)
34. T. Hannappel, S. Visbeck, L. Toben, F. Willig, Rev. Sci. Instrum. **75**, 1297–1304 (2004)
35. H. Döscher, S. Brückner, A. Dobrich, C. Höhn, P. Kleinschmidt, T. Hannappel, J. Cryst. Growth **315**, 10–15 (2011)
36. H. Döscher, S. Brückner, T. Hannappel, J. Cryst. Growth **318**, 563–569 (2011)
37. H. Döscher, P. Kleinschmidt, T. Hannappel, Appl. Surf. Sci. **257**, 574–580 (2010)
38. H. Döscher, A. Dobrich, S. Brückner, P. Kleinschmidt, T. Hannappel, Appl. Phys. Lett. **97**, 151905 (2010)
39. S. Brückner, H. Döscher, P. Kleinschmidt, O. Supplie, A. Dobrich, T. Hannappel, Phys. Rev. B. **86**, 195310 (2012)
40. D.S. Wuu, R.H. Horng, K.C. Huang, M.K. Lee, Appl. Phys. Lett. **54**, 236–238 (1989)
41. T.L. Barr, Appl. Surf. Sci. **15**, 1–35 (1983)
42. M.S. Hegde, R. Caracciolo, K.S. Hatton, J.B. Wachtman, Appl. Surf. Sci. **37**, 16–24 (1989)
43. R.I.G. Uhrberg, E. Landemark, Y.C. Chao, J. Electron Spectro. Relat. Phenom. **75**, 197–207 (1995)
44. M.L. Green, E.P. Gusev, R. Degraeve, E.L. Garfunkel, J. Appl. Phys. **90**, 2057–2121 (2001)
45. Z.H. Lu, M.J. Graham, D.T. Jiang, K.H. Tan, Appl. Phys. Lett. **63**, 2941–2943 (1993)
46. M. Niwano, H. Katakura, Y. Takeda, Y. Takakuwa, N. Miyamoto, A. Hiraiwa, K. Yagi, J. Vac. Sci. Technol. A-Vac. Surf. Films **9**, 195–200 (1991)
47. H. Hibino, M. Uematsu, Y. Watanabe, J. Appl. Phys. **100**, 8 (2006)
48. R. Tromp, G.W. Rubloff, P. Balk, F.K. Legoues, E.J. Vanloenen, Phys. Rev. Lett. **55**, 2332–2335 (1985)
49. K. Fujita, H. Watanabe, M. Ichikawa, J. Appl. Phys. **83**, 4091–4095 (1998)

50. Y. Wei, R.M. Wallace, A.C. Seabaugh, Appl. Phys. Lett. **69**, 1270–1272 (1996)
51. G.W. Rubloff, J. Vac. Sci. Technol. A-Vac. Surf. Films **8**, 1857–1863 (1990)
52. G.P. Tang, E. Peiner, H.H. Wehmann, A. Lubnow, G. Zwinge, A. Schlachetzki, J. Hergeth, J. Appl. Phys. **72**, 4366–4368 (1992)
53. S. Brückner, H. Döscher, P. Kleinschmidt, T. Hannappel, Appl. Phys. Lett. **98**, 211909 (2011)
54. Y. Kobayashi, Y. Shinoda, K. Sugii, Jpn. J. Appl. Phys. Part 1—Regul. Pap. Short Notes Rev. Pap. **29**, 1004–1008 (1990)
55. Y.J. Chabal, Surf. Sci. Rep. **8**, 211–357 (1988)
56. L. Mantese, U. Rossow, D.E. Aspnes, Appl. Surf. Sci. **107**, 35–41 (1996)
57. T.C. Shen, C. Wang, G.C. Abeln, J.R. Tucker, J.W. Lyding, P. Avouris, R.E. Walkup, Science **268**, 1590–1592 (1995)
58. R.A. Wolkow, Phys. Rev. Lett. **68**, 2636–2639 (1992)
59. R.J. Hamers, R.M. Tromp, J.E. Demuth, Phys. Rev. B. **34**, 5343–5357 (1986)
60. J.J. Boland, Adv. Phys. **42**, 129–171 (1993)
61. C.D. Wagner, L.E. Davis, M.V. Zeller, J.A. Taylor, R.H. Raymond, L.H. Gale, Surf. Interface Anal. **3**, 211–225 (1981)
62. S. Tougaard, Surf. Interface Anal. **11**, 453–472 (1988)
63. G. Hollinger, F.J. Himpsel, Appl. Phys. Lett. **44**, 93–95 (1984)
64. F.J. Grunthaner, P.J. Grunthaner, R.P. Vasquez, B.F. Lewis, J. Maserjian, A. Madhukar, Phys. Rev. Lett. **43**, 1683–1686 (1979)
65. K.O. Ng, D. Vanderbilt, Phys. Rev. B. **59**, 10132–10137 (1999)
66. H. Kageshima, K. Shiraishi, Phys. Rev. Lett. **81**, 5936–5939 (1998)
67. F.J. Himpsel, F.R. McFeely, A. Talebibrahimi, J.A. Yarmoff, G. Hollinger, Phys. Rev. B. **38**, 6084–6096 (1988)
68. M.P. Seah, S.J. Spencer, Surf. Interface Anal. **33**, 640–652 (2002)
69. T.A. Carlson, G.E. McGuire, J. Electron Spectrosc. Relat. Phenomena. **1**, 161–168 (1972/1973)
70. J.M. Hill, D.G. Royce, C.S. Fadley, L.F. Wagner, F.J. Grunthaner, Chem. Phys. Lett. **44**, 225–231 (1976)
71. Z.H. Lu, J.P. McCaffrey, B. Brar, G.D. Wilk, R.M. Wallace, L.C. Feldman, S.P. Tay, Appl. Phys. Lett. **71**, 2764–2766 (1997)
72. M.F. Hochella, A.H. Carim, Surf. Sci. **197**, L260–L268 (1988)
73. D.A. Cole, J.R. Shallenberger, S.W. Novak, R.L. Moore, M.J. Edgell, S.P. Smith, C.J. Hitzman, J.F. Kirchhoff, E. Principe, W. Nieveen, F.K. Huang, S. Biswas, R.J. Bleiler, K. Jones, J. Vac. Sci. Technol. B **18**, 440–444 (2000)
74. M. Sugo, Y. Takanashi, M.M. Aljassim, M. Yamaguchi, J. Appl. Phys. **68**, 540–547 (1990)
75. G. Zimmermann, A. Ougazzaden, A. Gloukhian, E.V.K. Rao, D. Delprat, A. Ramdane, A. Mircea, Mater. Sci. Eng. B-Solid State Mater. Adv. Technol. **44**, 37–40 (1997)
76. M. Dauelsberg, H. Hardtdegen, L. Kadinski, A. Kaluza, P. Kaufmann, J. Cryst. Growth **223**, 21–28 (2001)
77. H. Hardtdegen, M. Hollfelder, R. Meyer, R. Carius, H. Munder, S. Frohnhoff, D. Szynka, H. Luth, J. Cryst. Growth **124**, 420–426 (1992)
78. D. Keiper, P. Velling, W. Prost, M. Agethen, F.J. Tegude, G. Landgren, Jpn. J. Appl. Phys. Part 1 - Regul. Pap. Short Notes Rev. Pap. **39**, 6162–6165 (2000)
79. F. Habraken, A.E.T. Kuiper, Mater. Sci. Eng. R-Rep. **12**, 123–175 (1994)
80. K. Hirose, H. Nohira, K. Azuma, T. Hattori, Prog. Surf. Sci. **82**, 3–54 (2007)
81. J.A. Taylor, Appl. Surf. Sci. **7**, 168–184 (1981)
82. J.A. Taylor, G.M. Lancaster, J.W. Rabalais, J. Electron Spectro. Relat. Phenom. **13**, 435–444 (1978)
83. T.N. Wittberg, J.R. Hoenigman, W.E. Moddeman, C.R. Cothern, M.R. Gulett, J. Vac. Sci. Technol. **15**, 348–352 (1978)
84. R. van Weeren, E.A. Leone, S. Curran, L.C. Klein, S.C. Danforth, J. Am. Ceram. Soc. **77**, 2699–2702 (1994)

85. U. Gosele, T.Y. Tan, Appl. Phys. a-Mater. Sci. Proc. **28**, 79–92 (1982)
86. K. Miki, K. Sakamoto, T. Sakamoto, Surf. Sci. **406**, 312–327 (1998)
87. S.A. McQuaid, M.J. Binns, C.A. Londos, J.H. Tucker, A.R. Brown, R.C. Newman, J. Appl. Phys. **77**, 1427–1442 (1995)
88. J.E. Griffith, G.P. Kochanski, J.A. Kubby, P.E. Wierenga, J. Vac. Sci. Technol. A-Vac. Surf. Films **7**, 1914–1918 (1989)
89. J.J. Lander, J. Morrison, J. Chem. Phys. **37**, 729 (1962)
90. K. Heinz, Rep. Prog. Phys. **58**, 637–704 (1995)
91. O.L. Alerhand, A.N. Berker, J.D. Joannopoulos, D. Vanderbilt, R.J. Hamers, J.E. Demuth, Phys. Rev. Lett. **64**, 2406–2409 (1990)
92. F.A. Reboredo, S.B. Zhang, A. Zunger, Phys. Rev. B. **63**, 125316 (2001)
93. R.J. Hamers, P. Avouris, F. Bozso, Phys. Rev. Lett. **59**, 2071–2074 (1987)
94. Y.K. Sun, D.J. Bonser, T. Engel, Phys. Rev. B. **43**, 14309–14312 (1991)
95. T. Hannappel, S. Visbeck, M. Zorn, J.T. Zettler, F. Willig, J. Cryst. Growth **221**, 124–128 (2000)
96. N.J. Harrick, Phys. Rev. Lett. **4**, 224–226 (1960)
97. Y.J. Chabal, G.S. Higashi, S.B. Christman, Phys. Rev. B. **28**, 4472–4479 (1983)
98. Y.J. Chabal, Surf. Sci. **168**, 594–608 (1986)
99. Y.J. Chabal, J. Vac. Sci. Technol. A-Vac. Surf. Films **3**, 1448–1451 (1985)
100. J.J. Boland, Phys. Rev. Lett. **67**, 1539–1542 (1991)
101. N. Witkowski, R. Coustel, O. Pluchery, Y. Borensztein, Surf. Sci. **600**, 5142–5149 (2006)
102. R. Shioda, J. van der Weide, Appl. Surf. Sci. **130–132**, 266–270 (1998)
103. I. Nemeth, B. Kunert, W. Stolz, K. Volz, J. Cryst. Growth **310**, 1595–1601 (2008)
104. Y. Yanase, H. Horie, Y. Oka, M. Sano, S. Sumita, T. Shigematsu, J. Electrochem. Soc. **141**, 3259–3263 (1994)
105. S. Jeong, A. Oshiyama, Phys. Rev. Lett. **81**, 5366–5369 (1998)
106. L. Zhong, A. Hojo, Y. Matsushita, Y. Aiba, K. Hayashi, R. Takeda, H. Shirai, H. Saito, J. Matsushita, J. Yoshikawa, Phys. Rev. B. **54**, R2304–R2307 (1996)
107. Y. Kumagai, K. Namba, T. Komeda, Y. Nishioka, J. Vac. Sci. Technol. A-Vac. Surf. Films **16**, 1775–1778 (1998)
108. K. Kitahara, O. Ueda, Jpn. J. Appl. Phys. Part 2—Lett **33**, 1571–1573 (1994)
109. P. Bedrossian, T. Klitsner, Phys. Rev. Lett. **68**, 646–649 (1992)
110. A. Dobrich, P. Kleinschmidt, H. Doscher, T. Hannappel, J. Vac. Sci. Technol. B. **29**, 04D114 (2011)

Chapter 4
GaP(100) and InP(100) Surfaces

4.1 State-of-the-Art: Reconstructions of MOVPE-Prepared Surfaces

Both GaP and InP have been important materials in electronic or optoelectronic applications for several decades [1–3]. Recently, the GaP(100) surface has gained renewed attention as part of an almost lattice matched III–V compound on a silicon substrate [4, 5], while InP(100) has been employed as a substrate material for a low-band gap tandem solar cell [6]. In MOVPE environment, the well-known GaP(100) surface reconstructions [7] include a mixed (Ga-P) dimer surface for Ga-rich conditions, and a P-rich surface which has often been referred to as (2 × 1)/(2 × 2) surface. The InP(100) surface reconstructs in the same manner as evidenced by several theoretical [8] and experimental [9] studies. P-rich surfaces can only be obtained in the presence of hydrogen (available as a carrier gas in the MOVPE environment or as a precursor byproduct) [7, 9–11]. In the low energy electron diffraction (LEED) image these surfaces result in a (2 × 1) pattern with streaks at half order in the ×1 direction. According to Schmidt et al. [8] and Hahn et al. [12] this is due to a hydrogen-adsorbate structure, where the surface is terminated by buckled phosphorus dimers, each stabilized by a hydrogen atom which saturates one of the dangling bonds in the dimer [8, 9, 12].

The resulting structure is shown in Fig. 4.1: In a (2 × 2) cell, there are two phosphorus dimers in the top layer, which are oppositely buckled, with one hydrogen atom bonded to the 'down' atom of the P dimer. This leads to rows of buckled phosphorus dimers, where adjacent rows can be arranged in phase or out of phase. The in-phase arrangement results in a p(2 × 2) unit cell, while the out-of-phase arrangement corresponds to a c(4 × 2) unit cell, so a more accurate denotation of this surface reconstruction is p(2 × 2)/c(4 × 2), which we will adhere to in the following. A mixture of these surface reconstructions results in the above-mentioned LEED image. Schmidt et al. [8] have also calculated the corresponding STM image for InP(100), which has been experimentally observed by Vogt et al. [9]. Hahn et al. [12] provided calculations of the expected STM image

H. Döscher, *GaP heteroepitaxy on Si(100)*,
Springer Theses, DOI: 10.1007/978-3-319-02880-4_4,

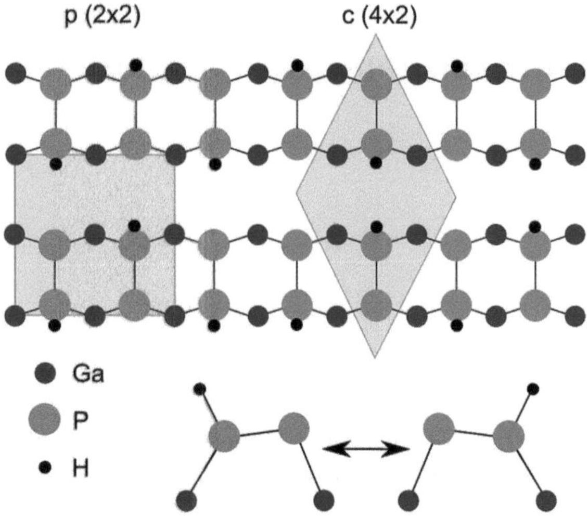

Fig. 4.1 *Top* view and *side* view of hydrogen stabilized GaP(100) and InP(100) surface structure after Hahn et al. [12] and Schmidt et al. [8]. The unit cells are indicated in the *top* view

of the p(2 × 2) structure of GaP(100), confirmed by the experiments of Töben et al. [7] and Kadotani et al. [10]. When imaging the filled states, a zigzag pattern is obtained, where the bright spots are due to the electron lone pair at the 'up' atom of the dimers.

Two very characteristic peaks in RA spectra of both the Ga-rich and the P-rich surface reconstructions of GaP(100) can be observed:

1. A pronounced negative peak below the energetic position of the interband transition E_0 [7]. This peak is only induced by electronic transitions of the surface Brillouin zone and appears at approximately the same energetic position for the Ga-rich and the P-rich surface reconstruction. The energetic shift between the respective peak position is just about 200 meV (see results, Fig. 4.5). However, the shape of these low energy peaks is different for the two different surface reconstructions (see below).
2. At the energetic position of the E_1 interband transition another, also pronounced, but positive peak arises for the P-rich surface reconstruction. It turns into a minimum, when the surface is changed towards the Ga-rich reconstruction which has been modeled and theoretically described by the (2 × 4) mixed dimer model [13].

Both peak shapes are quite sensitive to the sample temperature during measurement. When lowering the temperature, the peak intensity increases and the peaks sharpen significantly due to the reduced lattice vibrations and electron–phonon interactions [14]. Simultaneously, a blue shift of the energetic peak position occurs. The temperature-dependent energy shift of the peaks can be fitted with a model containing the Bose–Einstein occupation factor for phonons, as was shown for InP(100) [14], a surface which behaves very similar compared to GaP(100) [11, 15].

The principle origin of RAS signals is the optical anisotropy of the sample depending on a variety of different physical properties: For a single-crystalline material with cubic symmetry like GaP, only the surface is supposed to provide anisotropic features [16], if unintentional changes of the bulk symmetry such as local electric fields due to strain, doping [17] etc., can be neglected. Typically, the type of surface reconstruction determines the characteristic features of the corresponding RA spectra, but atomic order and sample temperature are decisive for the position and intensity of the observed peaks [18]. Beyond that, major signal contributions may arise from bulk-related anisotropies (such as ordering [19]) or from mesoscopic 3D-structures on the surface such as facets, steps, terraces or islands [20, 21].

The complexity of the different contributions to the RAS signal increases further when considering heteroepitaxy. On the one hand, strain and lattice changes such as ordering and defects may introduce bulk anisotropy in parts of such multilayer structures. On the other hand, optically accessible interior interfaces induce internal reflections, which interfere with the RAS measurement (Fabry–Perot oscillations) and which potentially contribute specifically to the RAS signal (interface anisotropy etc.) [4]. These contributions usually contain important characteristic information (such as layer thickness [22] and interface anisotropy [18]), but they superimpose during measurement. Various origins of the RAS signal have already been modeled by empirical [20] and semi-empirical [23] methods, some even by ab initio computational techniques [24], but still experimental RA spectra cannot reliably be predicted by theoretical calculations.

The idea to identify specific surface reconstructions with the corresponding RA spectra [25] generated high interest in optical in situ characterization. In the present case, therefore, the question arises, how well the theoretical RA spectrum has to fit the actual experimental spectrum for unambiguous identification of a surface reconstruction with a specific atomic model, e.g., the mixed dimer model of the Ga-rich GaP(100) surface.

Figure 4.2 shows a comparison of the experimental and the theoretical RA spectra which are currently available [7, 11, 12, 24, 26–29]. Although the calculation agree fairly well with the experimental data, a direct identification of surfaces is not possible via this route. In consequence, changing RAS features in a certain experiment might indicate a transition of the surface reconstruction, but can also be caused by variations in the order of the surface or just by a change of the sample temperature. Hence, the quantitative evaluation requires the precise knowledge and careful consideration of all contributions to the signal in order to exclude misinterpretations.

A quantitative interpretation of in situ RA spectra was already previously applied to determine intermediate states between the well-know P-rich (2 × 2)/c(4 × 2) and the III rich (2 × 4) reconstructions of the differently reconstructed (100) surfaces of GaP and InP [28, 30]. This approach was based on the assumption of partial surface area transitions due to changes in the reconstruction and will be discussed below. Another example for the quantitative evaluation of RA spectra is the desired in situ determination of anti-phase disorder in polar on

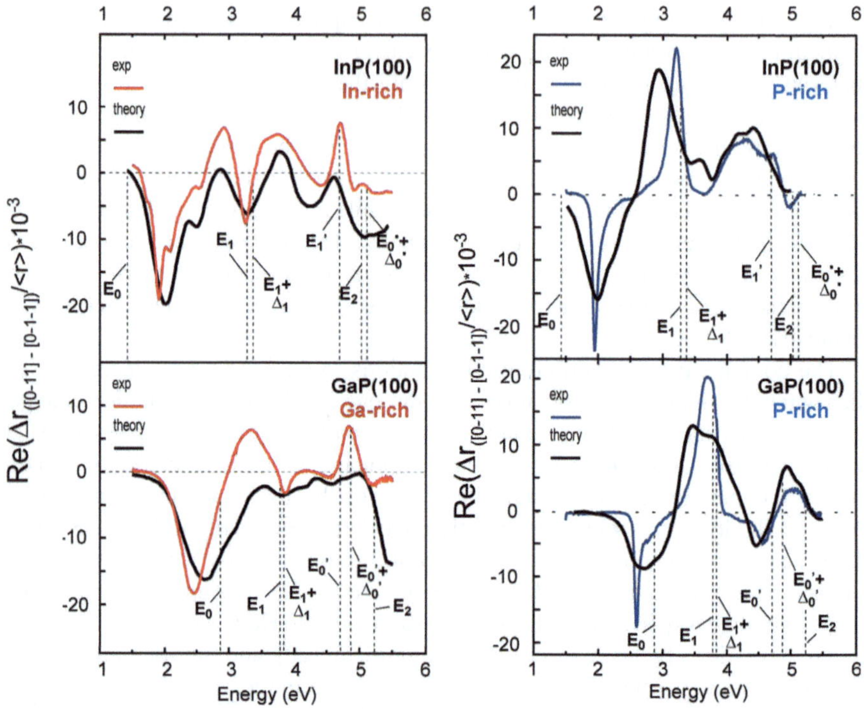

Fig. 4.2 Comparison of theoretical RA spectra for III-rich [54] and P-rich [8, 12] surfaces of InP(100) and GaP(100) to experimental data obtained at 20 K (After [55])

non-polar epitaxy of semiconductor materials. The deposition of GaP on Si(100) is considered as an adequate model system since pseudomorphic layers of relevant thicknesses (up to 40 nm) can be grown without relaxation [4, 5, 31]. Such samples allow the investigation of the principal impact of such heterointerfaces on the quality of the bulk material without any influence of lattice relaxation. Here, the formation of so-called anti-phase domains (APD) is a key challenge [5], since their internal boundaries (APB) contain structural defects in large quantities.

In principle, RAS provides a potential for the in situ characterization of III–V/Si(100) anti-phase disorder [4]. However, the quantitative evaluation of RA spectra requires detailed knowledge about the individual influences on the in situ RAS signatures of epitaxial compound semiconductor surfaces and the associated surface reconstruction mechanisms. The present study reports on the reproducible preparation of the atomically well-ordered, hydrogen terminated P-rich (2 × 2)/c(4 × 2) surface reconstructions of MOVPE prepared GaP(100) and InP(100). The impact of changes of the sample temperature as well as the reconstruction and the atomic order of the surfaces on the RAS signal were considered separately [32]. Although GaP(100) and InP(100) surfaces behave very similar, in particular during MOVPE preparation using hydrogen as the standard process gas, we found and characterized an additional surface reconstruction regime for nitrogen preparation of GaP(100), which does not exist for InP(100) [33]. For reference, the atomic structure of the P-rich reconstructed GaP(100) was characterized by STM. The images resolved rows of alternating buckled H-stabilized P-dimers arranged either in phase or out of phase, which refer to (2 × 2) and c(4 × 2) domains of the surface reconstruction, respectively. The investigations revealed a flipping mechanism [34], which switches the orientation of individual dimers, may transform the reconstruction domains into each other, and could also be found on P-rich prepared InP(100).

4.2 In Situ Control of GaP(100) Surfaces During MOVPE Preparation

In previous studies extensive surface sensitive characterization experiments via STM, LEED, and photoemission have been performed in our group, which were directly correlated to the RA spectra taken on the same GaP(100) samples [7, 11, 26, 35]. Thereby, in situ RAS allowed the identification of the basic GaP(100) surface reconstructions already during MOVPE preparation. Standard preparation conditions, which included hydrogen carrier gas and surface stabilization by TBP supply at elevated temperatures, led to the well known P-rich (2 × 2)/c(4 × 2) surface reconstruction [12]. If TBP supply was omitted during annealing, a transition to the also well known (2 × 4) reconstructed Ga-rich configuration of the surface occurred [7, 26]. However, the detailed shape of the RA spectra and the observed RAS intensity strongly depended on the preparation procedure before and the sample environment during the measurement, respectively. In the experiment associated with Fig. 4.3, a freshly prepared homoepitaxial GaP(100) surface was cooled down to 300 °C under TBP supply and afterwards the temperature was ramped up very slowly to 700 °C, while the P stabilization remained switched off and RA spectra were measured continuously. Here, basically all the relevant changes in the RA spectra were observed simultaneously.

The color-coded results of the described experiment in Fig. 4.3a clearly show the almost abrupt transition between the P- rich and Ga-rich reconstruction as soon as the temperature exceeded a critical value of about 490 °C (marked A) due to the preferential evaporation temperature of phosphorus from GaP [36]. The very slow increase of the temperature during the experiment led to quasi-equilibrium conditions, while each single spectrum was recorded. Selected spectra represent the P-rich and the Ga-rich surface configurations as well as an intermediate state in between (Fig. 4.3b). Their principal features agree very well with those observed in earlier studies [7, 11]. The color-coded plot of Fig. 4.3 also reveals continuous spectral changes due to a change of the sample temperature during the measurement: On the one hand, there is a red shift of spectral features with increasing temperature (marked B) according to the well-known shift of the bulk critical point

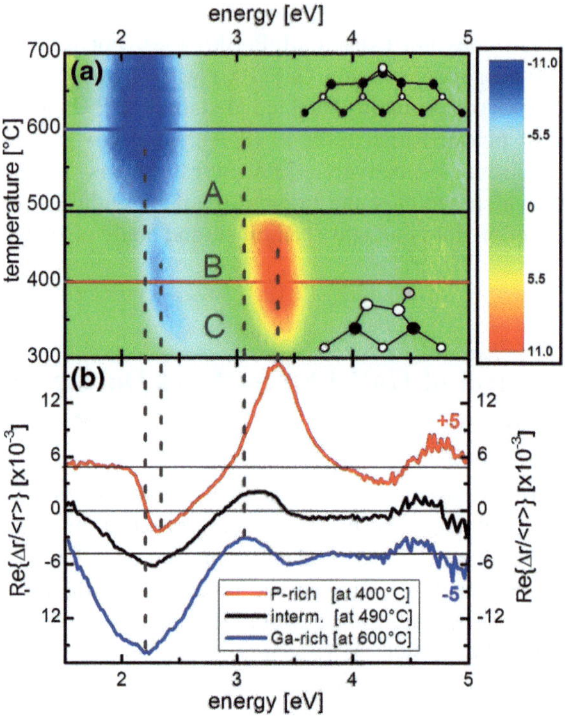

Fig. 4.3 **a** Color-coded representation of a continuous in situ RAS measurement (*colorplot*) at a GaP(100) surface while the sample temperature was ramped up from 300 to 700 °C at a rate of 1.5 K/min without TBP supply. Between 300 and 490 °C, spectra associated with a P-rich surface reconstruction were observed, while beyond 500 °C, a Ga-rich reconstruction appeared. In a narrow range around 490–500 °C, RA spectra of an intermediate surface reconstruction were taken. Letters mark regions where the development of the RA signal is predominantly caused by changes of the surface reconstruction (*A*), of the measurement temperature (*B*) and of the atomic order of the surface (*C*), respectively. **b** RA spectra representing two well-defined surface reconstructions, the P-rich (2 × 2)/c(4 × 2) at around 400 °C (*upper curve*) and the Ga-rich (2 × 4) at around 600 °C (*lower curve*), were extracted, and also the RA spectrum of the intermediate surface (*middle*) at ca. 490 °C. *Dashed vertical lines* at the peak positions refer to the respective features in the colorplot (From [32])

transitions, and simultaneously a broadening of the peaks due to increasing lattice vibrations and electron–phonon interactions, as was already observed and described for InP(100) [14].

On the other hand, in the temperature range from 300 to about 370 °C, there is a significant increase of the peak intensity (marked C) due to the desorption of excess phosphorus from the surface in analogy to previous findings drawn already from studies on InP [37, 38].

4.2.1 Transition from the P-rich to the Ga-rich Surface Reconstruction

The results of Fig. 4.3 revealed three independent parameters that changed the RAS signal of GaP(100) and superimposed during measurement: surface reconstruction, temperature and desorption of disordered excess phosphorus. Since we intended to apply the RA spectra for quantitative evaluations, we designed specific experiments to separate the impact of these parameter changes and to estimate their potential contribution on intensity and shape of the RA spectra. Usually phosphides are prepared with a high V/III ratio, i.e. with a high phosphorus partial pressure in comparison to the partial pressure of the III-valent component.

In contrast to the P-rich surface preparation, the parameter window for growing a III-rich surface termination, i.e. at a low V/III ratio, is very small. Hence, here we focused our studies on the P-rich GaP(100) surface reconstruction, which represents the standard surface for preparation in the MOVPE environment. Despite the critical preferential desorption temperature of about 490 °C [36] according to Fig. 4.3, sufficient supply with the P precursor TBP allowed to maintain the P-rich reconstruction even above 490 °C due to the equilibrium of desorption and adsorption of P at the surface. After the TBP supply was stopped in Fig. 4.4 ($t = 0$ s), the transformation from the P-rich to the Ga-rich surface reconstruction occurred at a fixed temperature of 525 °C. At that constant temperature, i.e. in absence of thermal effects on the RA spectra, we followed the transformation between the surface reconstructions with high resolution of the spectral region around the low energy peak at about 2.4 eV (Fig. 4.4) which is specifically surface sensitive, as it is only related to electronic transitions from the surface Brillouin zone [24]. Besides the two distinct spectra of the P-rich and the Ga-rich surface, there is also an intermediate range with a different RA spectrum between ca. 500 and 600 s.

The color-coded representation of the data in Fig. 4.4a confirms the specific character of the intermediate RAS signal. Obviously, the intensity connected to the P-rich surface regime vanished noticeably before the new peak of the Ga-rich termination started to appear. The associated in situ RA spectra (Fig. 4.4b) reflect the scenario described above in detail: A rather sharp peak at about 2.3 eV (red spectrum) represents the atomically ordered P-rich surface reconstruction without excess phosphorus (marked A). Afterwards, the topmost P layer desorbed successively. The 'intermediate' RA spectrum (spectrum at 500 s) during this transition already reflected the shape of the RA spectrum of the Ga-rich surface reconstruction that appeared in the further course. Notably, the spectrum of the Ga-rich state peaked at the slightly lower photon energy of about 2.2 eV (marked B) compared to the P-rich one. Its broader peak structure also contained considerable signal intensity at e.g. 2.0 eV (marked C), where the signal of the P-rich state (spectrum at 0 s) was almost absent.

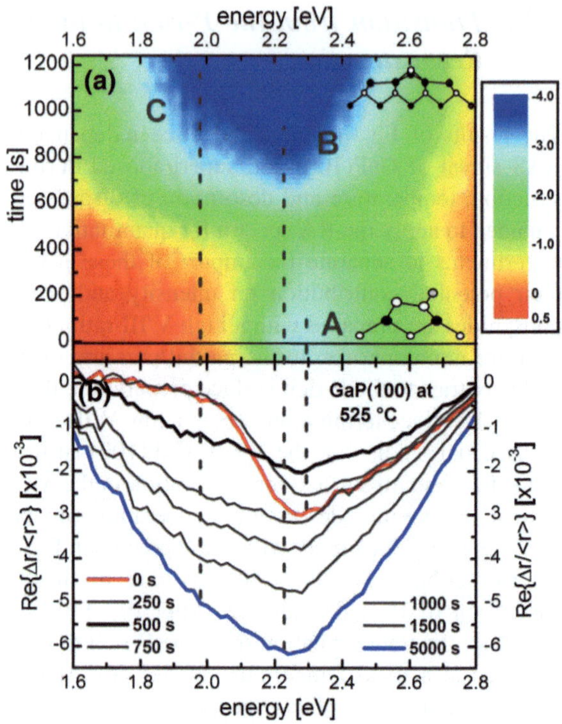

Fig. 4.4 **a** In situ RAS colorplot of a GaP(100) surface at the constant temperature of 525 °C after switch-off of the TBP supply. The measurement focused on the range of the low energy RAS peak at high resolution. The successive transformation from the P-rich to the Ga-rich surface reconstruction is accompanied with specific changes in the peak structure marked by the letters *A*, *B* and *C*. **b** RA spectra extracted from the colorplot at different times. *Dashed vertical lines* refer to the features observed in the colorplot correlate (From [32])

4.2.2 Thermal Development of RA Spectra of GaP(100)

Our experiments regarding the transition of the surface reconstruction from P-rich to the Ga-rich GaP(100) led to first conclusions for the quantitative RAS evaluation. With respect to post growth analysis, the P-rich state had to be sustained by adequate stabilization employing TBP when exceeding the critical temperature for preferential P desorption of about 490 °C [36] to prevent the loss of phosphorus. These results already show that the temperature dependence and the general impact of temperature induce changes in the RAS signal of P-rich GaP(100) and had to be studied and discussed in detail: In Fig. 4.3, we already derived three interrelated general trends of the RAS peak development upon the increase of temperature: a red shift of the peak position, a broadening of the peaks and a decrease of peak intensities. Figure 4.5 shows the data of the P-rich surface reconstruction cooled down from the preparation regime over ambient conditions to the cryogenic temperature of −250 °C (left side) as well as of the P-rich, the Ga-rich and the intermediate surface states at −250 °C in direct comparison (right side). The RAS peak in the low energy range of the spectrum was particularly well suited to investigate the temperature influences at well-defined surface reconstructions, as it is specifically surface sensitive.

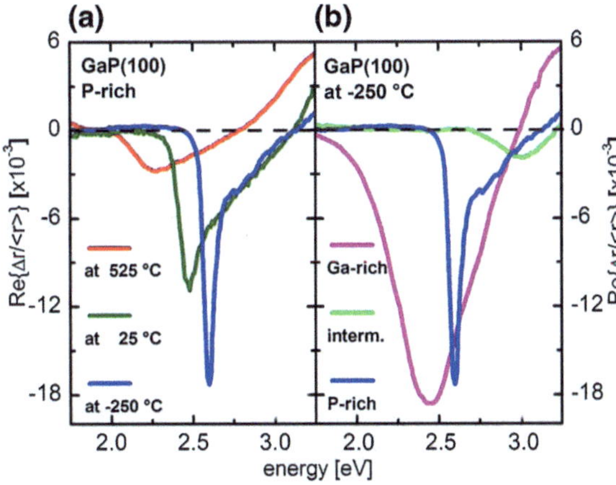

Fig. 4.5 Low energy RAS peak of **a** the atomically ordered, P-rich, (2 × 2)/c(4 × 2) GaP(100) surface reconstruction at different temperatures and **b** of the P-rich and of the Ga-rich reconstructions as well as of the intermediate GaP(100) surface, all measured at the cryogenic temperature of −250 °C for higher spectral resolution (From [32])

The evolution of the RA spectra in Fig. 4.5a upon cooling permitted several insights: The described general trends (sharpening of the spectrum and increase of the intensity) persisted over the entire temperature range under investigation. An impressive sharpening and increase of the low energy RAS peak occurred in Fig. 4.5a, corresponding to the energetic shift of bulk critical point transitions [39] and reduced lattice vibrations [40]. Analogous effects were already reported for the similar InP(100) surface [14]. The significantly reduced broadening revealed the peak structures in far more detail. While the broadened RA spectra at higher temperatures could tempt one to attribute these peaks to the identical origin, the comparison between the different RA spectra and peak positions of the two distinct surface reconstructions of GaP(100) at −250 °C (Fig. 4.5b) proved the different nature of the peaks. On the one hand, Fig. 4.5b clearly showed the complete disappearance of the low energy peak at around 2.5 eV in the intermediate preparation range, i.e. preparation conditions right between those for the P-rich and the Ga-rich termination. On the other hand, the P-rich case featured a significantly stronger sharpening of the RAS peak than the Ga-rich surface. Both reconstructions still showed comparable experimental RAS peak intensities in the low temperature spectra, but different peak areas and slightly different energetic positions. Hence, we found a systematic temperature dependence of GaP(100) RAS peaks concerning position, width and intensity in qualitative terms, but the behavior remained less predictable in quantitative terms. Even sophisticated theoretical first-principle total energy calculations with self-energy corrections of Hahn et al. [12] reproduced the detailed features of the experimental low temperature RA spectra, in particular regarding the low energy peak, only fairly well.

The impact of these considerations on our subsequent quantitative RAS analysis was diverse: The shape of our RA spectra shows quite clearly that the RA spectra displayed in Refs. [28, 30] contain strong interference artifacts, in particular in the range of the low energy peaks. Figure 4.5 also shows that it is ambiguous to superpose high temperature RA spectra of different surface reconstructions for the development of an atomic model [28] or for the evaluation of an unknown composition of surface reconstructions [30].

4.2.3 Excess Phosphorus Accumulation on the P-rich Surface

The P-terminated GaP(100) surface prepared via MOVPE with TBP supply right before or even during RAS measurements (Fig. 4.4) can result either in an atomically ordered P-rich surface or in a GaP(100) surface with excess phosphorus or even incompletely decomposed TBP on top. In our experiments the supply with TBP was essential to obtain and to sustain the P-rich surface reconstruction, in particular at elevated temperature ($T > 490$ °C). These two different surface terminations, i.e. the atomically ordered $(2 \times 2)/c(4 \times 2)$ P-rich GaP(100) surface reconstruction described by Hahn et al. [12] and the surface with excess phosphorus on top, have already been studied for InP(100) [38] and display also for GaP(100) significantly different RA spectra as Fig. 4.6 shows. The induced equilibrium state between dissociation of TBP as well as adsorption and desorption of P can include certain quantities of excess P atoms [7, 9] or even TBP fragments that mainly reduced the observed signal intensity.

At lower temperature ($T <$ ca. 370 °C), excess phosphorus on the surface was quite stable even without TBP supply. It desorbed gradually with rising temperature and/or the progress of time leading to a change of the RAS signal as observed in the experiment shown in Fig. 4.6. Here, in detail, an excess P layer was generated on a P-rich reconstructed GaP(100) surface by cooling the sample down to 275 °C under TBP supply. Immediately before it was switched off, an RA spectrum was taken (Fig. 4.6, curve with lowest intensity). After stopping the TBP supply the RAS signal changed successively (Fig. 4.6, curves with increasing intensities) and reflected the expected phosphorus desorption with the progress of time by an increase of the intensity only (at the high energy peak around 3.5 eV) or by an additional shift of the energetic position (low energy peak).

Short annealing cycles with gradually increased temperatures accelerated the excess phosphorus desorption significantly. To ensure directly comparable RA spectra, cooling down to 275 °C followed each annealing cycle. Via this procedure we separated the impact of the excess phosphorus on top of the surface on the RA spectra precisely from thermal effects. The increase of the signal intensity within this procedure stopped at around 365 °C (curve with highest intensity), where a spectrum almost identical to the subsequent cycle (385 °C) was measured. This indicated the complete desorption of the excess P from the surface associated with the atomically ordered, $(2 \times 2)/c(4 \times 2)$ P-rich GaP(100) surface reconstruction.

The in situ RAS investigation of excess phosphorus desorption (Fig. 4.6) stressed the different characteristics of the two major peaks. While the structure

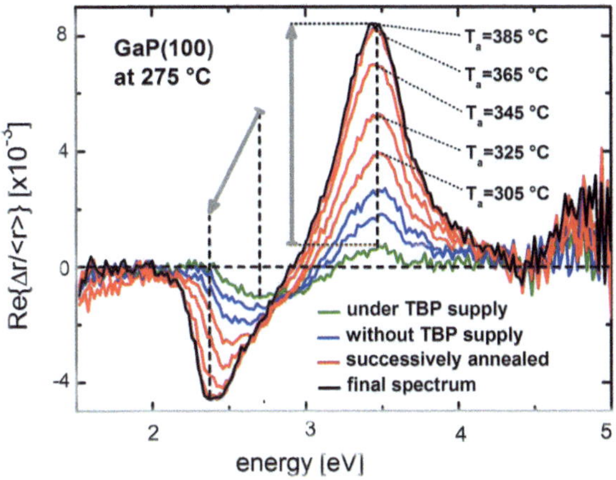

Fig. 4.6 RAS spectra of a GaP(100) sample, all taken at 275 °C, when successively preparing the atomically ordered (2 × 2)/c(4 × 2) surface reconstruction. The first spectrum was taken with TBP supply (lowest intensity), followed by two successive spectra just without TBP supply (increasing intensities). Then, the sample was annealed for short times at successively increasing temperatures and cooled down to 275 °C for the measurement of the individual RA spectra ($T_a > 300$ °C) until reaching the final RA spectrum of the P-rich surface (highest intensity) (From [32])

around the maximum at about 3.5 eV just scaled up at a constant energetic peak position with successive P desorption, the minimum at around 2.5 eV behaved differently. Here, the intensity increase accompanied a red shift of the low energy peak position. It gradually moved about 300 meV from 2.6 eV back to 2.3 eV, which represents the atomically ordered P-rich GaP(100) after desorption of excess phosphorus at that temperature. Hence, in Fig. 4.6 the excess phosphorus on the surface only reduced the intensity of the high energy peak in the range of the E_1 bulk critical point transition [11], but actually changed the energetic position of the low energy peak which is induced by transitions between specific surface states [14]. These obviously interacted with the excess P as the different behavior of the minimum implied, which already contributed to previous experiments as well (Figs. 4.3, 4.4): A red shift that was stronger than just the thermal shift was already visible in the color-coded GaP(100) surface evolution plot in the temperature range, where excess phosphorus desorbed (Fig. 4.3, T < ca. 370 °C). Also in Fig. 4.4 (T = 525 °C) at around t = 0 s a small, but immediate red shift of about 40 meV occurred after the TBP supply stopped and the excess phosphorus thus vanished from the surface.

The potential of excess phosphorus accumulation on P-rich GaP(100) surfaces implied restrictions, which have to be considered, when employing in situ RAS for the quantitative analysis. In these experiments (Figs. 4.3, 4.4 and 4.6), we determined a suitable and large parameter window for the preparation of the atomically ordered P-rich GaP(100) surface regarding the relevant temperatures: The excess phosphorus upon P-rich surfaces was only stable at T < 370 °C (Fig. 4.6), while the atomically ordered P-rich reconstruction itself persisted without P supply until 490 °C (Figs. 4.3, 4.4). Within these temperatures, the preparation of the ordered P-rich surface reconstruction just required the interruption of TBP stabilization and an adequate annealing period. Regarding the similar scenario for the closely related surfaces of InP(100), the parameter window for the preparation of the ordered P-rich surface is different and much smaller, since the preferential desorption temperature of phosphorus from the bulk is for InP much lower than for GaP(370 vs. 490 °C) [36, 41]. Consequently, the narrow preparation window necessitated a sophisticated procedure to achieve the ordered P-rich InP(100) surface reconstruction [37, 38].

Once achieved, the P-rich GaP(100) reconstruction did not change at all, when cooling down towards low temperatures (Fig. 4.5). The RAS investigations of Fig. 4.4 (below 0 s) indicated that GaP(100) under TBP stabilization at higher temperatures forms an only slightly modified P-rich surface structure with few excess phosphorus on top [9] (Fig. 4.4). In consequence, the P-rich reconstruction of a homoepitaxial GaP(100) surface defined a suitable reference for quantitative RAS analysis [4]: It featured a well-defined atomic structure as well as an intense and well-established RA spectrum. We described a simple procedure for the reproducible preparation of the atomically ordered P-rich GaP(100) surface and presented methods to measure reliable reference RA spectra at different temperatures, explicitly including the established MOVPE regimes for GaP growth.

4.3 GaP(100) and InP(100) Preparation in an Alternative Process Ambient

Nitrogen is an established alternative carrier gas for MOVPE processing [42–45]. Since we observed Si_3N_4 formation during Si(100) preparation (Sect. 3.3.5), N_2 process gas cannot be considered as an overall inert alternative. However, standard MOVPE processing involves significantly lower temperatures, for which molecular nitrogen is certainly expected to be non-reactive. In the following, we use surface sensitive in situ optical spectroscopy for a comparison of GaP(100) and InP(100) preparation by N_2-based MOVPE processes to the established results obtained using H_2 standard process gas.

Figure 4.7 depicts a comparison of in situ RAS colorplots measured during comprehensive GaP(100) surface preparation experiments using either N_2 (a) or H_2 (b) as MOVPE carrier gas. In both cases, freshly prepared homoepitaxial GaP (100) surfaces from a standard MOVPE process were cooled down to 300 °C under TBP supply. Subsequently, the temperature was ramped up very slowly to 700 °C under flow of the respective carrier gas, while the P stabilization remained switched off and RA spectra were measured continuously. The very slow increase of the temperature during the experiment led to quasi-equilibrium conditions,

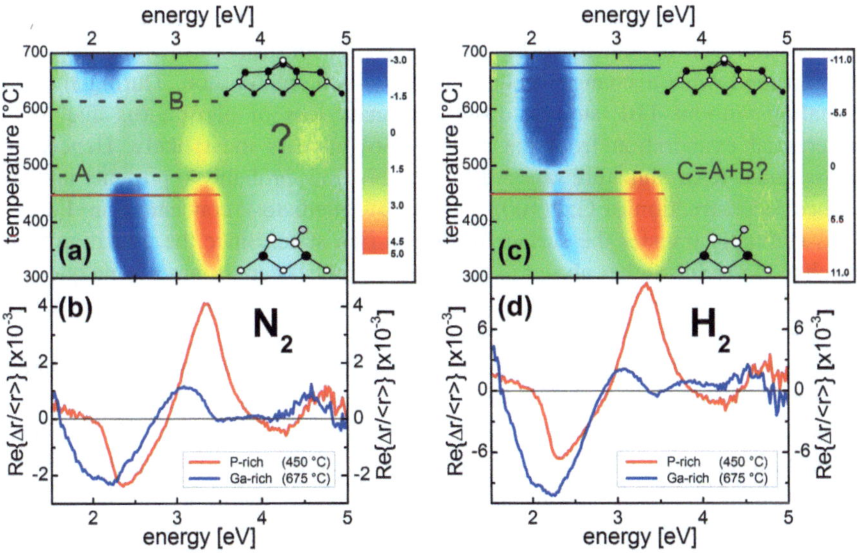

Fig. 4.7 Comparison of the MOVPE-preparation of GaP(100) surfaces in nitrogen (**a**, **b**) and hydrogen (**c**, **d**) by means of color-coded in situ RAS results (**a**, **c**) measured while the sample temperature was ramped up slowly (at a rate of 1.5 K/min) from 300 to 700 °C without TBP supply (After [33])

while each single spectrum was recorded. The color-coded representation of the measured RAS data in Fig. 4.7 characterizes all GaP(100) surface structures that can be prepared in the given environments. Since the H_2 data have already been described in detail in Sect. 4.2, our description focuses the results obtained under N_2 flow, and, in particular, the differences in the observation.

Major parts of the in situ RAS colorplots in Fig. 4.7 appear very similar. Actually, the shape of the selected spectra (b vs. d) identifies the characteristic P-rich (red) and Ga-rich (blue) surface reconstructions of GaP(100) at lower (450 °C) and higher (675 °C) process temperatures, respectively. In contrast to the situation in the standard process ambient (using hydrogen), where a direct transition from the P-rich to the Ga-rich state of the surface occurs (c), a wide intermediate region with a different RAS signature was detected when using nitrogen as alternative carrier gas (a). The dashed horizontal lines in Fig. 4.7a, c indicate temperatures at which characteristic changes of the RAS signal occurred. Similar to GaP(100) in H_2, the P-rich surface reconstruction remained stable up to a temperature of about 490 °C under N_2 flow, where the reconstruction eventually changes. Instead of the almost abrupt transition to the characteristic RAS signal of the Ga-rich mixed dimer termination associated to the use of hydrogen (c), during nitrogen preparation (a), the RAS signal seems to disappear for a more than 100 K wide temperature regime, until the Ga-rich features finally start to develop at about 620 °C.

Significant changes in the RAS signature of a compound semiconductor surface usually indicate a transition of the surface reconstruction, which typically accompanied with a change of the surface stoichiometry, too. During thermal annealing without TBP supply, disproportionate loss of phosphorus due to its preferential evaporation from the surface [36] appears most likely. Hence, the intermediate RAS signature observed under N_2 flow probably represents another surface reconstruction of GaP(100) with an intermediate P concentration. Further P losses finally transfer the surface to the Ga-rich state. Since the transition from P-rich to Ga-rich is rather abrupt in the standard process environment, the question arises if the presence of hydrogen enhances the removal of phosphorus in a way that the intermediate surface state can not be formed.

To derive more detailed information from the GaP(100) surface preparation experiment in nitrogen shown in Fig. 4.7, we conducted a detailed assessment of the associated spectroscopic data depicted in Fig. 4.8 and in Fig. 4.9 in combination with relevant LEED pattern obtained by contamination free transfer to UHV for benchmarking the in situ RAS spectra [33]. In the range of 460–495 °C, we observed the continuous decay of the P-rich surface reconstruction identified with the characteristic RA spectrum and a (2 × 1)-like LEED pattern (Fig. 4.8, red). At 495 °C, the first RAS peak disappeared completely, while the second peak and the relative minimum at about 4 eV became less pronounced (Fig. 4.8, turquoise). This state of the GaP(100) is associated with a diffuse diffraction pattern with only weak (1 × 1) reflexes. During further annealing at continuously increasing temperature (Fig. 4.9, green), the first RAS peak remains absent until about 600 °C

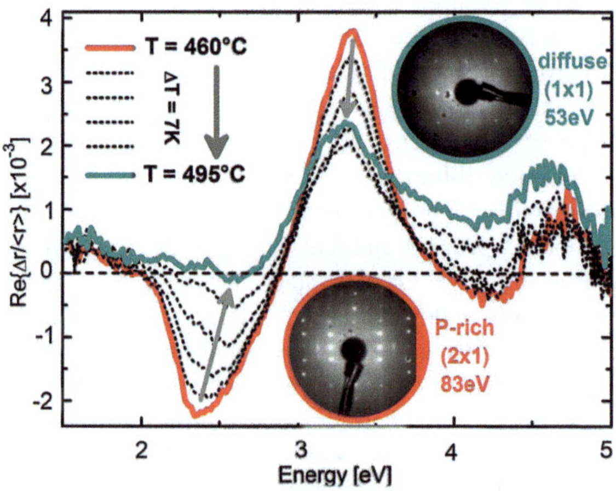

Fig. 4.8 Detailed development of GaP(100) in situ RA spectra during continuous annealing in nitrogen showing the decay of features associated with the initially P-rich surface between 460 and 495 °C in steps of 7 K. The insets show LEED patterns of the well-known (2 × 1)-like surface reconstruction (*red*, at 83 eV) and of diffuse (1 × 1) diffraction associated with the intermediate state of the surface (*turquoise*, at 53 eV) (From [33])

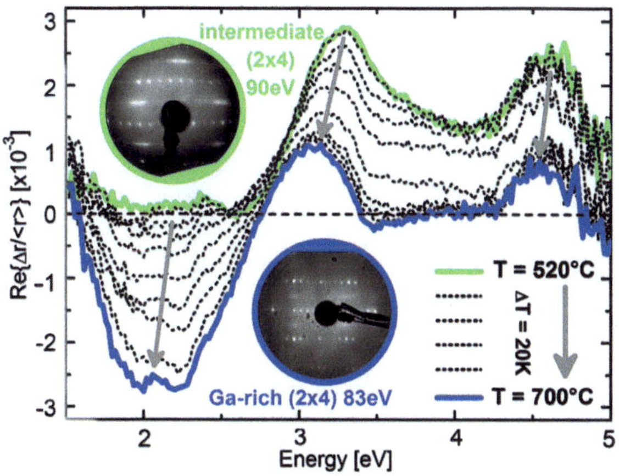

Fig. 4.9 Detailed development of GaP(100) in situ RA spectra during continuous annealing in nitrogen showing the successive formation of features associated with the Ga-rich surface between 520 and 700 °C in steps of 20 K. The insets both show LEED patterns with (2 × 4) symmetry, one measured after annealing in N_2 at 520 °C (*green*, 90 eV), the other after standard preparation of the Ga-rich surface reconstruction (*blue*, 83 eV) (From [33])

while the intensity of the second peak structure stabilizes at a slightly increased level compared to the value observed at about 490 °C. Accordingly, we observed some slightly yellow phase at about 3.3 eV in the colorplot of Fig. 4.7, which already was associated with a (2 × 4) LEED pattern (Fig. 4.9, green).

Above 600 °C the characteristic RAS features of the Ga-rich surface reconstruction GaP(100) started to develop (Fig. 4.9, blue), which are attributed (2 × 4) diffraction pattern, too. Since LEED just indicates the symmetry of the surface reconstruction, different atomic configurations potentially feature identical patterns, as we suppose for N_2 prepared GaP(100) according to the undisputable difference in the shape and intensity of the RA spectra.

In general, the development of GaP(100) and InP(100) is very similar, in particular during standard MOVPE preparation, and virtually identical in a qualitative sense. The atomic models of the established P-rich (2 × 2)/c(4 × 2) and III-rich (2 × 4) mixed dimer surface reconstructions (Fig. 4.11c) are replaceable, and abrupt, direct transitions between both terminations are observed during annealing in hydrogen. Hence, we also conducted InP(100) surface preparation experiments in nitrogen (Fig. 4.10) to check for the existence of an additional intermediate surface reconstruction regime in analogy to the GaP(100) results.

Figure 4.10 depicts a comparison of in situ RAS colorplots measured during comprehensive InP(100) surface preparation experiments using either N_2 (a) or H_2 (b) as MOVPE carrier gas. In analogy to the GaP(100) experiments shown in Fig. 4.7, continuous RAS measurement monitored the development of freshly prepared InP(100) surfaces, while the temperature was ramped up very slowly from 300 to 480 °C without TBP supply under flow of the respective carrier gases.

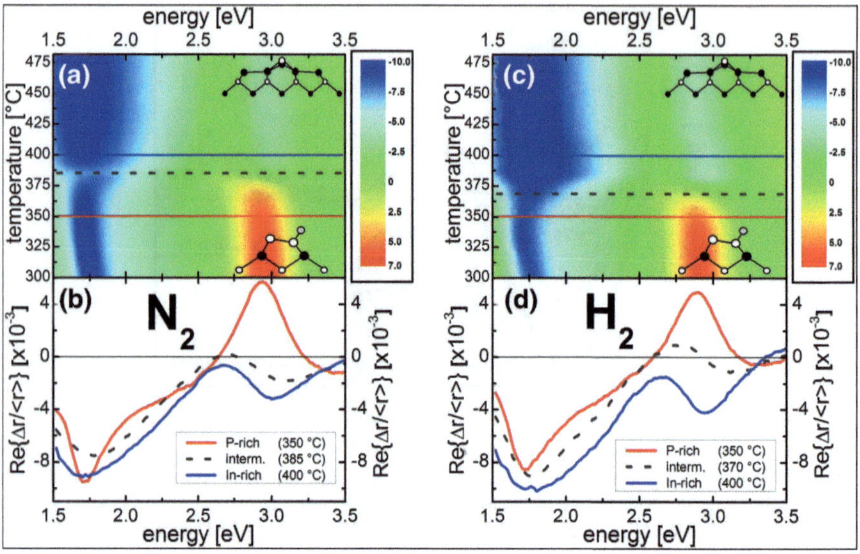

Fig. 4.10 Comparison of the MOVPE-preparation of InP(100) surfaces in nitrogen (**a**, **b**) and hydrogen (**c**, **d**) by means of color-coded in situ RAS results (**a**, **c**) measured while the sample temperature was ramped up slowly from 300 to 480 °C without TBP supply (From [33])

In contrast to the results obtained with GaP(100) (Fig. 4.7), both InP(100) colorplots depicted in Fig. 4.10 (a vs. c) show a similar, direct transition from the initial V-rich to the final III-rich surface reconstruction independent of using alternative (a) or standard (c) carrier gas. In detail, some differences in the transition region (around 375 °C) can be recognized. Selected spectra (b vs. d) identify the characteristic RAS features associated with P-rich (red) and In-rich (blue) InP(100) at lower (350 °C) and higher (400 °C) process temperatures, respectively. Only in N_2, the intermediate spectrum (dashed grey) confirms a saddle point of the intensity of the first peak.

Nevertheless, the comparison of GaP(100) and InP(100) surface preparation in nitrogen revealed significant qualitative differences in the observed surface transformation processes, which can be explained according to the theoretical determined surface phase diagrams of the hydrogen exposed GaP(100) [12] and InP(100) [8] depicted in Fig. 4.11a and b, respectively. Both diagrams identify the theoretical stability of the different reconstructions in dependence of chemical potentials of group III $\Delta\mu$(Ga/In) and hydrogen $\Delta\mu$(H) atoms on the surfaces. The established P-rich and Ga-rich reconstructions common for standard MOVPE preparation under H_2 flow are marked red and green, respectively, while green color identifies different phases with (2 × 4) symmetry and higher phosphorus content compared to the Ga-rich mixed dimer model. In detail, Fig. 4.11c shows the established [8, 46] atomic models for these surface reconstructions ordered with decreasing phosphorus content. Starting from the hydrogen-stabilized P-rich (2 × 2)-2D-2H (left), the top layer P concentration in the surface unit cell decreases

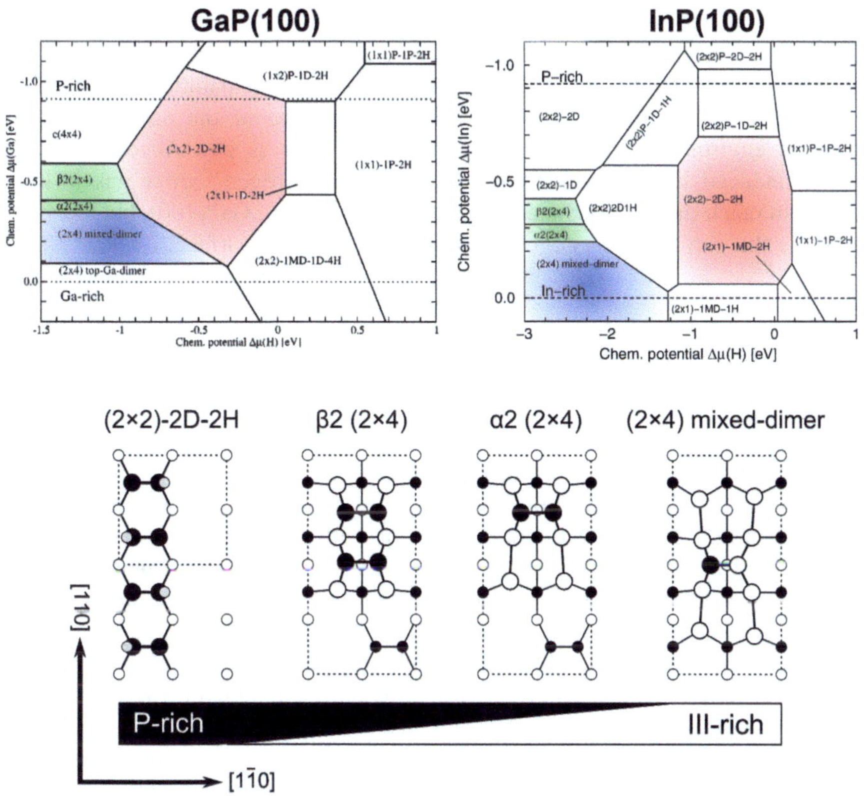

Fig. 4.11 Comparison between the calculated phase diagram of the hydrogen exposed GaP(100) (**a**) and InP(100) surfaces after Refs. [12] and [8], where *broken lines* indicate thermodynamic limitations of the group III chemical potentials and $\Delta\mu(H)$ is given with respect to molecular hydrogen. Colored areas identify surface reconstructions relevant for this study: the P-rich configuration typical for MOVPE preparation (*red*), the III-rich mixed-dimer surfaces achievable by annealing without TBP supply (*blue*), and (2 × 4) reconstructions with intermediated phosphorus content (*green*). (**c**) Top view of the relevant (100) surface reconstruction models ordered with decreasing phosphorus content starting from the hydrogen-stabilized P-rich (2 × 2)-2D-2H (*left*), over the intermediate $\beta2$ (2 × 4) and $\alpha2$ (2 × 4) to the finally group-III-rich (2 × 4) mixed-dimer surface (*right*). According to Refs. [8, 46], *filled* (*empty*) *circles* represent phosphorus (group-III) atoms and the atomic positions in the uppermost two layers are identified by larger symbols (From [33])

from 1 monolayer (ML) over 1/2 and 1/4 ML for the intermediate $\beta2$ (2 × 4) and $\alpha2$ (2 × 4) structures to 1/8 ML for the group-III-rich (2 × 4) mixed-dimer surface (right), respectively. In parallel, the group-III content increases stepwise from 0 over 3/4 ML to finally 1 1/8 ML in the mixed-dimer structure.

The comparison between surface phase diagrams of Fig. 4.11 shows adjacent regimes for the P-rich and the different (2 × 4) phases of GaP(100), while the (2 × 4) terminations of InP(100) are well separated from its P-rich phase. In particular, the regime of the $\beta 2$ (2 × 4) and $\alpha 2$ (2 × 4) structures of InP(100) appear at a significantly lower chemical potential of hydrogen, compared to the moderate values assigned to GaP(100). During MOVPE preparation, the chemical potentials are determined by the surface temperature of the sample and the chemical environment defined by the choice of process gas. Accordingly, the change induced by the switch from hydrogen to nitrogen carrier gas is large enough to reach the intermediate surface reconstruction of GaP(100), but too small to bridge the gap to the analogous phase of the InP(100) surface.

4.4 Atomic Resolution of the P-rich Surfaces and Dynamical Observation

Figure 4.12 shows an image of one of the GaP(100) surfaces at room temperature after transfer to UHV. The image quality compares favorably with the images presented in refs. [7, 10]. A direct comparison with the calculated images [12] is possible, and in region A of Fig. 4.12 the zigzag pattern of the calculated filled states image is clearly visible with only the 'up' atoms contributing to the image. As mentioned above, this corresponds to the p(2 × 2) reconstruction. In region B the honeycomb structure is observed, which corresponds to the c(4 × 2) reconstruction. The regions of these reconstructions only extend over a few unit cells, leading to numerous locations, where transitions between the two symmetries exist. This indicates that the additional energy required for neighboring dimers to buckle in the same direction and therefore to violate the symmetry of the two reconstructions is relatively low. In a few places, irregular bright structures are

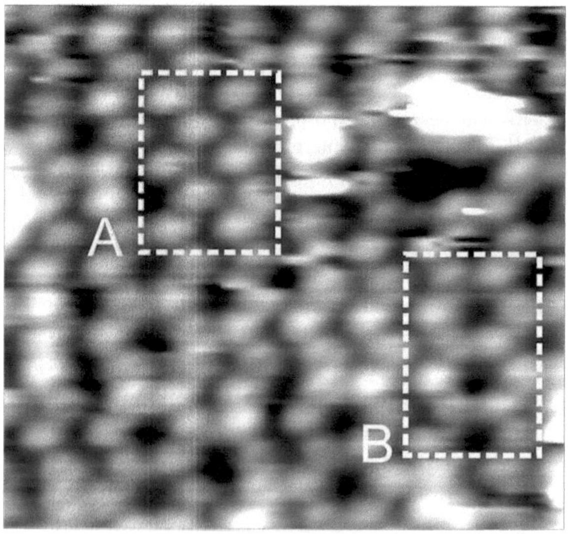

Fig. 4.12 Filled state image of GaP(100) surface at room temperature. Regions *A* and *B* exhibit p(2 × 2) and c(4 × 2) reconstructions, respectively. Scale 65 × 60 Å (From [34])

visible. These are likely to be due to residues from precursor fragments, which have not been removed completely by the annealing procedure, or due to missing H atoms or impurities.

The image of the GaP(100) surface in Fig. 4.12, which results directly from MOVPE preparation may be compared with images in Ref. [9] of the P-rich InP(100) surface which were obtained after an annealing step in UHV. Although the defect density is higher in Fig. 4.12, the similarity of the underlying surface structure is striking. Figure 4.12 confirms that this surface should be referred to as p(2 × 2)/c(4 × 2) reconstructed.

4.4.1 Evidence for Flip-flop Motion on P-rich GaP(100) Surfaces

The clean Si(100) surface possesses some similarity to this surface; in the case of (2 × 1)-reconstructed Si(100) a complete layer of dimers terminates the surface without involving hydrogen, forming a (2 × 1) reconstruction and leaving one electron in each dangling bond. These electrons form π-bonds, and the dimers tilt, with charge being transferred from the down atoms to the up atoms [47], giving rise to buckled dimers that exhibit p(2 × 2) and c(4 × 2) unit cells [48, 49]. However, STM images at room temperature show symmetric dimers [50], which is attributed to a flip-flop motion, where the dimers fluctuate between the two equilibrium positions at a rate too high to be resolved at room temperature, and only in the vicinity of defects or at sufficiently low temperature buckled dimers are observed in STM images. Clearly, in the case of GaP(100) and InP(100) the dimers do not appear symmetrically at room temperature. However, we investigated the possibility that they undergo a flip-flop motion at a much lower rate by comparing successive images of the surface.

Figure 4.13 shows a region of two successive images at room temperature for this comparison. The rectangular frames indicate positions, where the surface changes between the two images. In frame 1 there are six protrusions, of which the three on the left-hand side in the frame remain unchanged. The three protrusions on the right-hand side appear in a zigzag arrangement in Fig. 4.13a, while in Fig. 4.13b they are aligned vertically. In frame 2 the three protrusions on the right-hand side remain unchanged, while the zigzag arrangement of the three protrusions on the left-hand side is flipped with respect to a vertical axis. The process which is associated with the flipping of a protrusion with respect to a vertical axis becomes clear when one considers the atomic model according to Hahn et al. The protrusions in the images are due to the electron lone pair of the 'up' atom of the P-dimer. The flipping therefore implies that the 'up' atom shifts downwards and vice versa, so that the buckling of the dimer reverses (i.e. the dimer is mirrored). According to the atomic model this means that the H atom, which stabilizes the P-dimer, changes position from one P-atom in the dimer to the other. This is

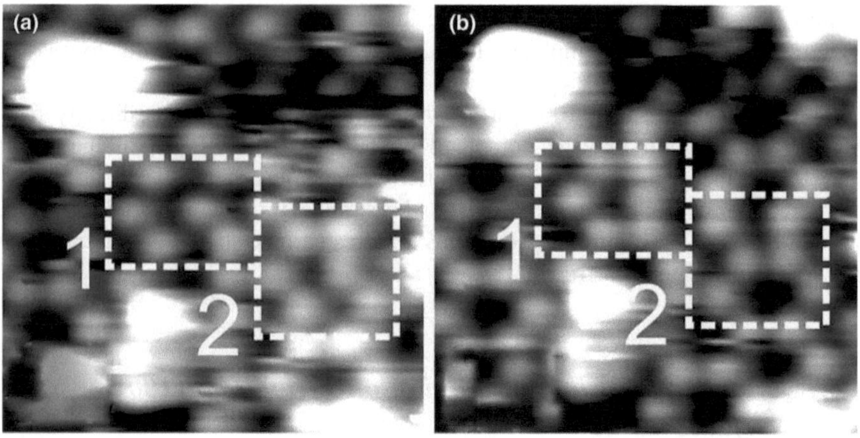

Fig. 4.13 Consecutive filled state images of GaP(100) surface at room temperature. Frames *1* and *2* indicate regions where dimer flipping occurs between the images. The time elapsed between images is about 1 min, while the time required to completely image one atom is about 1.5 s. Scale 45 × 45 Å (From [34])

distinctly different from the case of the Si(100) (2 × 1) surface, where the flip-flop motion only results from a charge transfer from one Si atom to the other and their related rehybridization.

4.4.2 Temperature Dependence of the Dimer Flipping Processes

One may speculate that the motion and change of the bonding arrangement of the H atom required in the case of the GaP(100) surface, is the reason for the comparatively low rate of the dimer flipping in this case. In order to investigate the possibility of a thermally activated process we carried out experiments at the lower sample temperature of 120 K (while the scanner is maintained at room temperature in our STM when working at low temperature).

Figure 4.14 shows sections of three successive images recorded at 120 K in which the rectangular frames highlight a region where dimer flipping can be observed. Between Fig. 4.14a and b two dimers belonging to adjacent rows flip, while from Fig. 4.14b to c they flip back again. Hence, even at 120 K the buckling of the dimers is not frozen. The sequence of the images shown, containing the dimers flipping back and forth, gives an indication of the rate at which the process occur. We can estimate an upper and lower limit of this rate based on the time to image one atom, during which its appearance does not change and on the number of changes between images, respectively. For both temperatures, these rates lie between 10 and 0.03 s^{-1}. While in the case of simple thermal activation of the

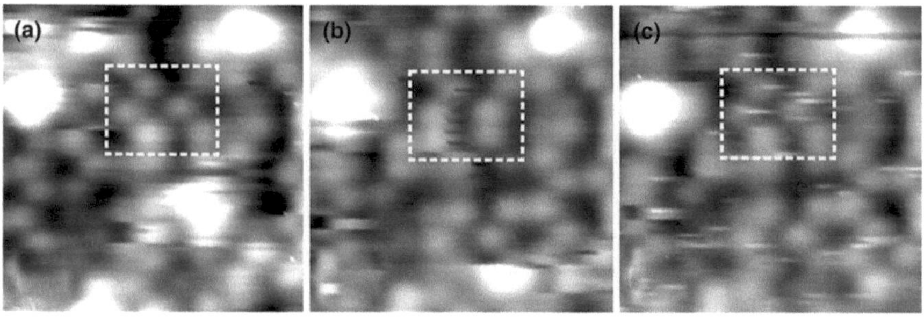

Fig. 4.14 Consecutive filled state images of GaP(100) surface at 120 K. The *frame* indicates a location where two dimers flip back and forth in the images. The time elapsed between consecutive images here is about 16 s, while the time required to completely image one atom is about 0.4 s. Scale 45 × 45 Å (From [34])

same system at room temperature and at 120 K a difference of many orders of magnitude in frequency should result, the images are inconsistent with a difference of significantly more than two orders of magnitude. The flip-flop motion seen in Fig. 4.14 should increase to a frequency at room temperature which would prevent resolution of the dimer in the STM image, in disagreement with the observation.

On the other hand, the energy required to flip a particular dimer may depend on the direction of the buckling of its neighbors and of the dimers in adjacent rows. The configuration of these dimers in the vicinity of the flipping event is different for the images at room temperature and at 120 K. In the image at 120 K the dimers flip between configurations which are energetically degenerate when considering nearest neighbors (in each configuration one of the two neighboring dimers is oriented in the same direction as the one that flips, while the other neighbor is oriented in the opposite direction). On an ideal surface, this situation occurs at the boundary between p(2 × 2) and c(4 × 2) symmetry, but in the images shown, the vicinity contains several defects. Still, the degeneracy may apply to first order to the dimers in question and may facilitate the observed flipping. In the images at room temperature, some dimers within one row flip individually, while in other locations several neighboring dimers are seen to flip together between images. This may be associated with different energy barriers. Based on these observations it is likely that the energy barrier governing the events observed at 120 K is different from those which apply to the events shown in the images at room temperature. Here, theoretical considerations might provide further clarification.

4.4.3 Dimer Flipping on P-rich InP(100) Surfaces

Dimer flipping processes have also been observed on InP(100) p(2 × 2)/c(4 × 2) surfaces [34]. Such a process is shown in Fig. 4.15 where two successively measured occupied states room temperature STM images are depicted. The arrows

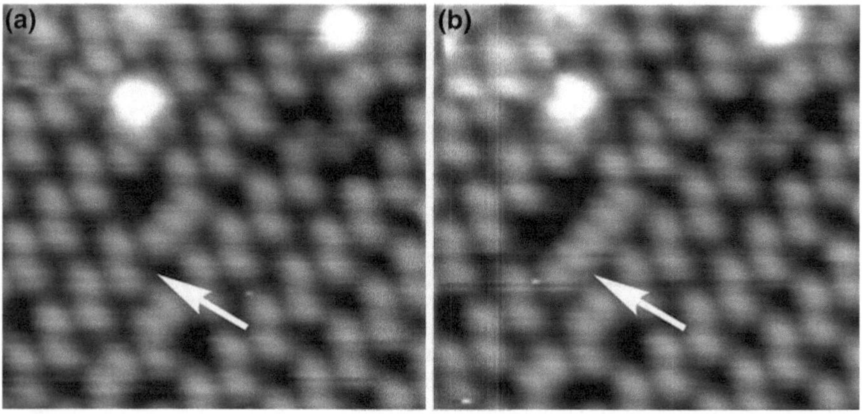

Fig. 4.15 STM observation of a dimer-flipping process at a InP(100) p(2 × 2)/c(4 × 2) surface unit (indicated by *arrows*) in two successive occupied states STM images measured at room temperature. Such flipping processes have been observed several times at different locations on the surface. Scale 58 × 58 Å, sample bias voltage: −5.2 V (From [34])

indicate the same zigzag unit in both micrographs. It can be seen that the bright protrusion of this unit is moved to the other side in the second image with respect to the first one. Such flipping processes have been observed several times at different locations on this surface. They take place in the vicinity of surface defects, such as voids or adsorbed species on the top layer (contaminations). A possible reason is that the flipping process is supported by the strain produced by the defects.

The dimer flipping supports the model of a H-stabilized phosphorus dimer. The calculations by Pulci et al. [29] and Schmidt et al. [8] have shown that unterminated P–P dimers at the InP(100) surface are inconsistent with a buckled dimer structure. Hence, the 2 × 1 cell at the surface must be due to a mixed dimer or an adsorbate at the P–P dimer. The dimer flipping contradicts the mixed dimer structure, as an exchange of different kinds of dimer atoms as the basis for the observed flipping is highly unlikely. As an adsorbate hydrogen is the only possible candidate since XPS measurements on these surfaces (not shown here) would have revealed the presence of other elements. Hence, the only structure in agreement with our experimental observation is the hydrogen stabilized P–P dimer.

Based on the temperature-dependent measurements, we conclude that, most likely, a tip-induced process takes place, comparable to other systems with tip-induced motion of surface atoms [51–53], which may be aided by surface strain due to defects in the vicinity. Possible mechanisms of tip-sample interaction include heating due to the high local current density or the influence of the high electric field caused by the STM tip. However, we cannot entirely rule out that activation energies of the involved configurations are sufficiently different to compensate for the difference in the available thermal energies between room temperature and 120 K. Theoretical studies could elucidate the relation between the configurations and the energies involved. A more detailed experimental analysis with a larger set of images under different conditions will be carried out to obtain data on a larger variety of atomic configurations and to study the interaction of tip and sample. In particular, we will investigate the dependence of the dimer flipping on bias voltage and tunneling current and whether some control of the observed effect by targeted local variation of the tunneling parameters can be obtained.

References

1. M. Gershenzon, R.A. Logan, D.F. Nelson, Phys. Rev. **149**, 580 (1966)
2. R.A. Logan, H.G. White, W. Wiegmann, Appl. Phys. Lett. **13**, 139 (1968)
3. K.J. Bachmann Annu, Rev. Mater. Sci. **11**, 441–484 (1981)
4. H. Döscher, T. Hannappel, B. Kunert, A. Beyer, K. Volz, W. Stolz Appl. Phys. Lett. **93**, 172110 (2008)
5. B. Kunert, I. Nemeth, S. Reinhard, K. Volz, W. Stolz, Thin Solid Films **517**, 140–143 (2008)
6. N. Szabo, B.E. Sagol, U. Seidel, K. Schwarzburg, T. Hannappel, Phys. Status Solidi-Rapid Res. Lett. **2**, 254–256 (2008)
7. L. Töben, T. Hannappel, K. Moller, H.J. Crawack, C. Pettenkofer, F. Willig, Surf. Sci. **494**, L755–L760 (2001)
8. W.G. Schmidt, P.H. Hahn, F. Bechstedt, N. Esser, P. Vogt, A. Wange, W. Richter, Phys. Rev. Lett. **90**, 4 (2003)
9. P. Vogt, T. Hannappel, S. Visbeck, K. Knorr, N. Esser, W. Richter, Phys. Rev. B **60**, R5117–R5120 (1999)
10. N. Kadotani, M. Shimomura, Y. Fukuda, Phys. Rev. B **70**, 5 (2004)
11. T. Hannappel, L. Toben, K. Moller, F. Willig, J. Electron. Mater. **30**, 1425–1428 (2001)
12. P.H. Hahn, W.G. Schmidt, F. Bechstedt, O. Pulci, R. Del Sole, Phys. Rev. B **68**, 033311 (2003)
13. W.G. Schmidt, F. Bechstedt, Surf. Sci. **409**, 474–484 (1998)
14. S. Visbeck, T. Hannappel, M. Zorn, J.T. Zettler, F. Willig, Phys. Rev. B. **63** (2001)
15. P.H. Hahn, W.G. Schmidt in Surface ordering of P-rich InP (001): Hydrogen stabilization versus electron correlation (World Scientific Publ Co Pte Ltd, 2002), pp.163–167
16. P. Weightman, D.S. Martin, R.J. Cole, T. Farrell, Rep. Prog. Phys. **68**, 1251–1341 (2005)
17. Z. Kollonitsch, K. Moller, H.J. Schimper, C. Giesen, M. Heuken, F. Willig, T. Hannappel, J. Cryst. Growth **261**, 289–293 (2004)
18. O. Hunderi, J.T. Zettler, K. Haberland, Thin Solid Films **472**, 261–269 (2005)
19. J.S. Luo, J.M. Olson, S.R. Kurtz, D.J. Arent, K.A. Bertness, M.E. Raikh, E.V. Tsiper, Phys. Rev. B **51**, 7603–7612 (1995)
20. O. Acher, S.M. Koch, F. Omnes, M. Defour, M. Razeghi, B. Drevillon, J. Appl. Phys. **68**, 3564–3577 (1990)
21. K. Möller, Z. Kollonitsch, C. Giesen, M. Heuken, F. Willig, T. Hannappel, J. Cryst. Growth **248**, 244–248 (2003)
22. J.T. Zettler, Prog. Cryst. Growth Charact. Mater. **35**, 27–98 (1997)
23. M. Murayama, K. Shiraishi, T. Nakayama, Jpn. J. Appl. Phys. Part 1—Regul. Pap. Short Notes Rev. Pap. **37**, 4109–4114 (1998)
24. W.G. Schmidt, N. Esser, A.M. Frisch, P. Vogt, J. Bernholc, F. Bechstedt, M. Zorn, T. Hannappel, S. Visbeck, F. Willig, W. Richter, Phys. Rev. B **61**, 16335–16338 (2000)
25. F. Bechstedt, O. Pulci, W.G. Schmidt, Phys. Status Solidi A **175**, 5–16 (1999)

26. N. Esser, W.G. Schmidt, J. Bernholc, A.M. Frisch, P. Vogt, M. Zorn, M. Pristovsek, W. Richter, F. Bechstedt, T. Hannappel, S. Visbeck, J. Vac. Sci. Technol. B **17**, 1691–1696 (1999)
27. A.M. Frisch, W.G. Schmidt, J. Bernholc, M. Pristovsek, N. Esser, W. Richter, Phys. Rev. B **60**, 2488–2494 (1999)
28. D.C. Law, Y. Sun, R.F. Hicks, J. Appl. Phys. **94**, 6175–6180 (2003)
29. O. Pulci, K. Ludge, P. Vogt, N. Esser, W.G. Schmidt, W. Richter, F. Bechstedt, Comput. Mater. Sci. **22**, 32–37 (2001)
30. M.J. Begarney, C.H. Li, D.C. Law, S.B. Visbeck, Y. Sun, R.F. Hicks, Appl. Phys. Lett. **78**, 55–57 (2001)
31. I. Nemeth, B. Kunert, W. Stolz, K. Volz, J. Cryst. Growth **310**, 1595–1601 (2008)
32. H. Döscher, T. Hannappel, J. Appl. Phys. **107**, 123–523 (2010)
33. H. Döscher, K. Möller, T. Hannappel, J. Cryst. Growth **318**, 372–378 (2011)
34. P. Kleinschmidt, H. Döscher, P. Vogt, T. Hannappel, Phys. Rev. B **83**, 155–316 (2011)
35. T. Hannappel, S. Visbeck, L. Toben, F. Willig, Rev. Sci. Instrum. **75**, 1297–1304 (2004)
36. C.T. Foxon, B.A. Joyce, R.F.C. Farrow, Griffith. Rm J. Phys. D-Appl. Phys. **7**, 2422–2435 (1974)
37. T. Hannappel, S. Visbeck, M. Zorn, J.T. Zettler, F. Willig, J. Cryst. Growth **221**, 124–128 (2000)
38. T. Hannappel, L. Toben, S. Visbeck, H.J. Crawack, C. Pettenkofer, F. Willig, Surf. Sci. **470**, L1–L6 (2000)
39. S. Zollner, M. Garriga, J. Kircher, J. Humlicek, M. Cardona, G. Neuhold, Thin Solid Films **233**, 185–188 (1993)
40. P. Lautenschlager, P.B. Allen, M. Cardona, Phys. Rev. B **31**, 2163–2171 (1985)
41. R.F.C. Farrow, J. Phys. D-Appl. Phys. **7**, 2436–2448 (1974)
42. G. Zimmermann, A. Ougazzaden, A. Gloukhian, E.V.K. Rao, D. Delprat, A. Ramdane, A. Mircea, Mater. Sci. Eng. B-Solid State Mater. Adv. Technol. **44**, 37–40 (1997)
43. M. Dauelsberg, H. Hardtdegen, L. Kadinski, A. Kaluza, P. Kaufmann, J. Cryst. Growth **223**, 21–28 (2001)
44. H. Hardtdegen, M. Hollfelder, R. Meyer, R. Carius, H. Munder, S. Frohnhoff, D. Szynka, H. Luth, J. Cryst. Growth **124**, 420–426 (1992)
45. D. Keiper, P. Velling, W. Prost, M. Agethen, F.J. Tegude, G. Landgren, Jpn. J. Appl. Phys. Part 1—Regul. Pap. Short Notes Rev. Pap. **39**, 6162–6165 (2000)
46. W.G. Schmidt, Appl. Phys. A Mater. Sci. Process. **75**, 89–99 (2002)
47. A. Ramstad, G. Brocks, P.J. Kelly, Phys. Rev. B **51**, 14504–14523 (1995)
48. R.A. Wolkow, Phys. Rev. Lett. **68**, 2636–2639 (1992)
49. H. Shigekawa, K. Miyake, M. Ishida, K. Hata, H. Oigawa, Y. Nannichi, R. Yoshizaki, A. Kawazu, T. Abe, T. Ozawa, T. Nagamura, Jpn. J. Appl. Phys. Part 2—Lett. **35**, L1081–L1084 (1996)
50. R.J. Hamers, R.M. Tromp, J.E. Demuth, Phys. Rev. B **34**, 5343–5357 (1986)
51. D.M. Eigler, E.K. Schweizer, Nature **344**, 524–526 (1990)
52. G. Meyer, B. Neu, K.H. Rieder, Appl. Phys. A Mater. Sci. Process. **60**, 343–345 (1995)
53. P. Ebert, M.G. Lagally, K. Urban, Phys. Rev. Lett. **70**, 1437–1440 (1993)
54. W.G. Schmidt, F. Bechstedt, J. Bernholc J. Vacuum Sci. Technol. B **18**, 2215–2223 (2000)
55. T. Hannappel, Habilitation (Freie Universität Berlin, 2005)

Chapter 5
GaP Growth on Si(100) and Anti-phase Disorder

5.1 State-of-the-Art: Polar on Non-polar Epitaxy and Anti-phase Disorder

The technological interest in the superior electronic and optoelectronic properties of III–V semiconductors promotes the research in heteroepitaxy on silicon (Si) substrates for integration in established micro-electronics. However, major challenges have to be met at the III–V/Si heterojunction: crucial issues are differences in lattice constants and thermal expansion coefficients as well as the formation of the heterovalent (polar-on-non-polar) interface necessitating a suitable substrate preparation prior to heteroepitaxy. New defect mechanisms—typically not observed in III–V homoepitaxy—arise from the interface with the Si(100) substrate and need to be controlled to achieve defect concentrations suitable for applications in advanced optoelectronic devices [1].

Due to the small lattice mismatch (0.37 % at 300 K), epitaxial growth of gallium phosphide (GaP) on Si(100) begins pseudomorphic. Therefore, the GaP/Si(100) system is an important model system for polar-on-non-polar epitaxy. Thin, pseudomorphic GaP films on Si(100) are particularly suitable for the study of the formation of the heterovalent interface and to characterize the influence of the interface on the evolution of defects in detail.

Anti-phase disorder in the deposited III–V epilayers has been identified as one of the major challenges [1, 2] and is subject of recent studies [3, 4]: the two-domain structure of the reconstructed Si(100)(2 × 1) surface is a main obstacle for the epitaxial growth of single-domain zinc blende structures of III–V semiconductors such as GaP (Fig. 1.3). As a consequence, undesired structures of anti-phase domains (APDs) separated by anti-phase boundaries (APBs) are formed, which typically propagate in growth direction. APDs are regions of anti-sites [5] with respect to each other which feature orthogonal orientations of their local lattices in the surface plane [6].

H. Döscher, *GaP heteroepitaxy on Si(100)*,
Springer Theses, DOI: 10.1007/978-3-319-02880-4_5,

5.1.1 Detection of Anti-phase Disorder: Established Techniques

The high similarity of the APDs with the main phase complicates the detection of anti-phase disorder. Since rather unfavorable bonds (III–III and/or V–V) are the specific feature of the APBs separating different domains, destructive methods such as anisotropic etching [7] or adequate post growth annealing [8] can enhance the topographic contrast of APB intersections with the surface (Fig. 5.1). In direct analogy to anisotropic etching, preferential GaP evaporation associated with weaker bonds in the APBs occurs if certain annealing conditions are applied [8]. Consequently, AFM images of such GaP/Si(100) samples visualize the resulting trenches which correspond to APBs as height contrast.

Figure 5.1 displays AFM images of identically prepared GaP films grown on Si(100) substrates with different misorientation of 0.1° (a), 2° (b), and 6° (c). To exhibit general surface morphology and anti-phase disorder at the same time, the selected height scale does not resolve the morphology around APBs, consequently the comparatively small APDs appear completely black. Instead, the images provide a good impression of the achieved surface quality in main phase areas, where the observed step structure is consistent with misorientation of the respective Si(100) substrates.

Direct evidence for APDs as a bulk feature can be gained from X-ray scattering [1]. Due to the reversal of the stacking sequence in APDs compared to the main phase, the APBs, which are the junctions of both phases, are visible in high-resolution TEM images [9] as shown in Fig. 5.2. All images show APBs in thin heteroepitaxial GaP films grown on slightly misoriented Si(100) substrates. In principle, the APBs propagate in growth direction, but either straight (a) or kinked towards tilted lattice planes (b). Since the specimen thickness is on the order of

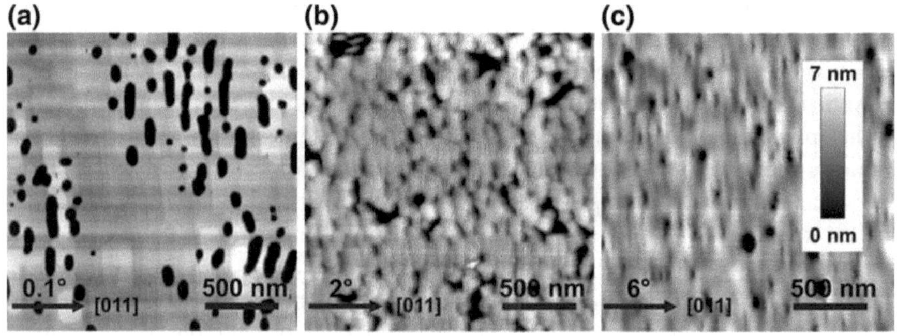

Fig. 5.1 AFM images of the surface of thin heteroepitaxial GaP films grown by MOVPE on Si(100) substrates with different off-cut angles of 0.1° (**a**), 2° (**b**), and 6° (**c**). For the visualization of anti-phase disorder, specific annealing conditions induced preferential desorption at APBs prior to AFM measurement. Due to the choice of contrast for better visibility of surface steps, the complete APD areas appear *dark* in all images (**a–c**)

several tens of nm, kinking from the viewing direction leads to a split or widened appearance of the APBs (c).

Even direct resolution of APDs and their geometrical distribution can be achieved by TEM if proper imaging conditions are applied [10]. Both cross sectional (Fig. 5.3) and plan-view (Fig. 5.4) studies yielded characteristic APD contrast by specific dark-field imaging combined with adequate tilts of the specimen as described by Ref. [4] for GaP/Si(100). Figure 5.3 shows different cross-sectional dark-field TEM images of different heteroepitaxial GaP/Si(100), where anti-phase disorder is highlighted by bright/dark contrast. The applied imaging conditions decide, whether either the main phase (a) or the APDs (b) appears bright.

The higher magnification image of Fig. 5.3c shows a single anti-phase domain and its boundaries to the matrix of the main phase, which typically appear darker than any of both phases. Hence, the different character of APBs is also visible with dark-field contrast: While the APB on the right is depicted as a thin straight line in direct analogy to Fig. 5.2a, the APB on the left is slightly tilted in some part and most notably much thicker than the other one. In analogy to Fig. 5.2c, kinking from the viewing direction explained this blurred appearance.

Figure 5.4 shows two plan-view dark-field TEM images of a GaP/Si(100) sample with slightly different contrast conditions. While in part (a) both phases equally appear grey and anti-phase disorder is visualized by darker network of APBs, part (b) shows distinct dark/bright contrast between the APDs and the main phase. As already seen in the cross-sectional dark-field TEM images of Fig. 5.3, the thickness of the APBs in the images can be applied as an approximate measure of their kinking

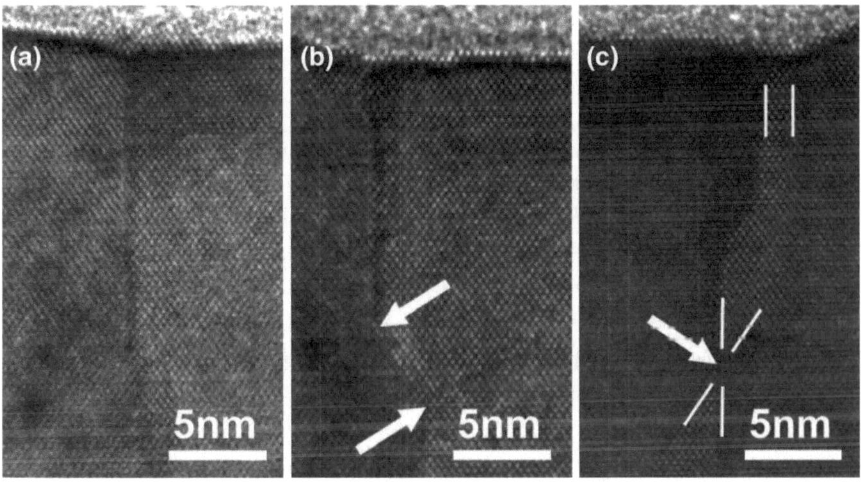

Fig. 5.2 High-resolution TEM images displaying APBs obtained at cross-section specimens prepared from MOVPE-grown GaP films on Si(100) substrates with 0.1 misorientation. In general, APBs propagate in growth direction (**a**), but do not necessarily grow exactly straight (**b**). APB configurations which are not perpendicular to the image plane can appear twice (**c**)

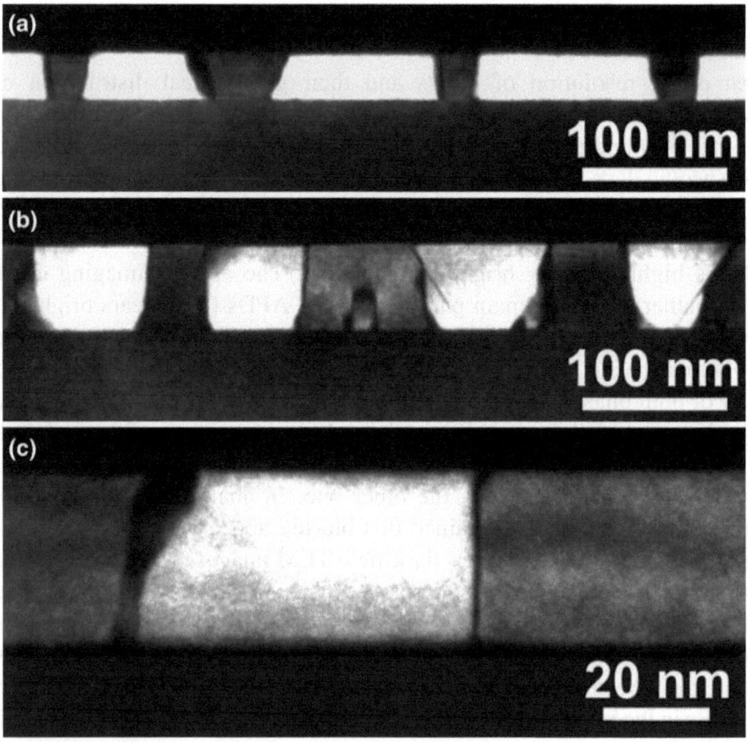

Fig. 5.3 Cross-section TEM images with specific *dark-field* contrast for visualization of anti-phase disorder. According to the choice of imaging conditions, either the main phase (**a**) or the APDs (**b**) can appear bright. Magnification of a single APD (**c**) shows the difference between the APBs at both edges

relative to the viewing direction. In the case of plan-view image, both lateral directions of the surface are observed simultaneously. The comparison between typical APB widths and APD sizes allows a very rough estimate of the probability APB self-annihilation at the growth conditions applied during sample preparation.

5.1.2 *Correlation of Anti-phase Disorder and Surface Structure*

The reconstructions of the GaP(100) surface are well-studied. A P-rich surface reconstruction of GaP(100) is typically obtained after growth by metal–organic vapor phase epitaxy (MOVPE) [11, 12]. In MOVPE, hydrogen from the process gas stabilizes the buckled surface P-dimers [13] with alternating tilts [Fig. 5.5 (inset)]. In analogy to the reconstructions of the clean Si(100) surface [14], the

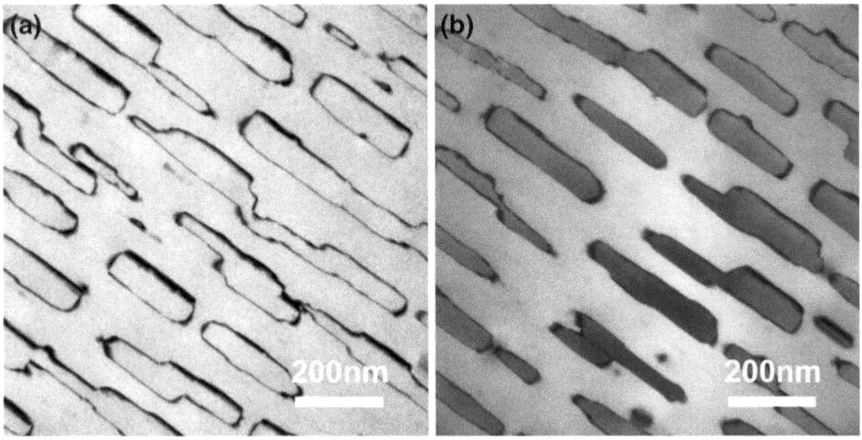

Fig. 5.4 Plan-view TEM images with specific *dark-field* contrast for visualization of anti-phase disorder. Depending on the applied imaging conditions, either APBs appear with *dark contrast* (**a**) or *dark/bright* contrast between APDs and the main phase may be achieved (**b**)

arrangement of parallel or anti-parallel dimer rows on P-rich GaP(100) leads to the coexistence of (2 × 2) and c(4 × 2) structures at the surface [Fig. 5.5 (left side)]. In low energy electron diffraction (LEED) this surface yields a (2 × 1) superstructure pattern with additional streaks [11]. The P-rich GaP(100) surface reconstruction is also associated with an intense characteristic RA spectrum which enables reliable in situ control of the surface preparation [6].

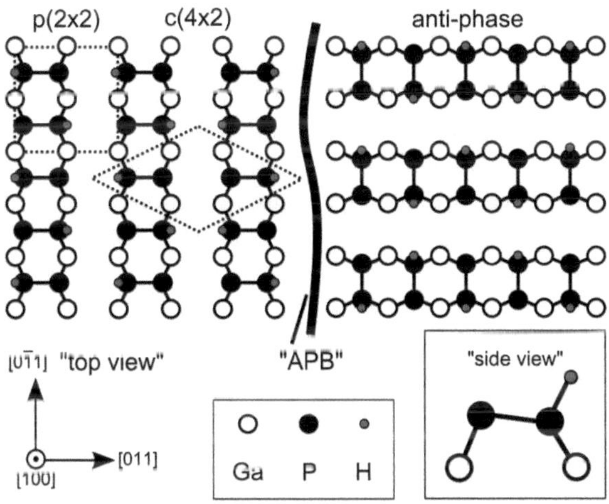

Fig. 5.5 Dimer model of the P-rich GaP(100) surface reconstruction: based on buckled P-dimers stabilized by a single hydrogen atom (see *inset*), the surface is arranged in parallel and anti-parallel dimer rows resulting in a combined p(2 × 2)/c(4 × 2) symmetry (*left*). Only on hetero-epitaxial GaP/Si(100) surfaces, also perpendicular dimer orientation occurs (*right*) due to anti-phase disorder (From [15])

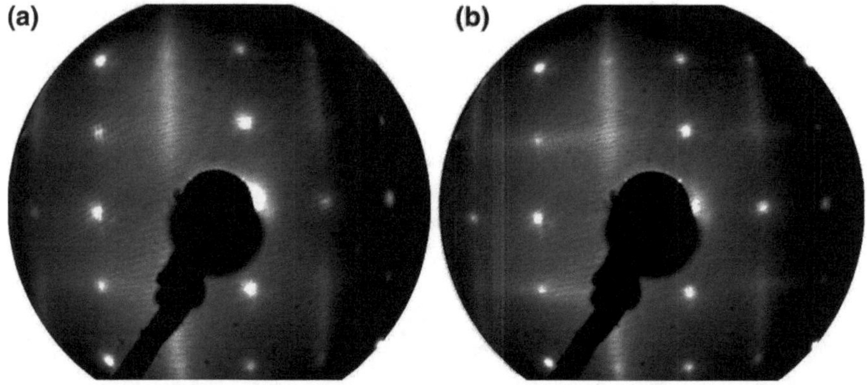

Fig. 5.6 Typical 60 eV LEED patterns measured at heteroepitaxial GaP films grown by MOVPE on 2° off-cut Si(100) substrates. The single-domain surface (**a**) exactly reproduces the (2 × 1)-like pattern expected for P-rich prepared GaP(100), while the image of the distinctly two-domain surface (**b**) contains significant signal of the mutually perpendicular surface reconstruction domain as an indication for the anti-phase disorder in the GaP film

In principle, the identical reconstruction mechanism is expected for P-rich prepared surfaces of heteroepitaxial GaP/Si(100) samples. However, Fig. 5.5 shows that the P-dimers align in mutually perpendicular directions reflecting the anti-phase disorder in the film. LEED measurements laterally average over the microscopic domain structure. In general, we expect a combined (2 × 1)/(1 × 2) diffraction pattern (Fig. 5.6) for a GaP film grown on Si(100), in which the intensity of the partial patterns is determined by the distribution of APDs.

Figure 5.6 shows a comparison of two LEED pattern measured at the P-rich prepared surfaces of two GaP/Si(100) samples. While the left image (a) appears basically identical to the (2x1)-like LEED pattern of a P-rich prepared homo-epitaxial GaP(100) surface (Fig. 4.8, inset), the right one (b) shows streaks in both surface directions and additional half-order diffraction spots.

In direct analogy to the LEED observations a two-domain reconstructed Si(100) (Fig. 3.11), the measurement laterally integrates the domain distribution of the surface. Accordingly, the sample associated with Fig. 5.6b showed significant anti-phase disorder at its surface, while the LEED pattern of Fig. 5.6a does not indicate the presence of (1 × 2)-like minority surface reconstruction domains. However, it appears questionable to what extent LEED observations can confirm the single-domain character of a heteroepitaxial III–V/Si(100) surface.

The characteristic RAS features associated with the P-rich GaP(100) surface are essentially determined by the surface reconstruction [6]. According to the definition of RAS (Eq. 5.1), the measured anisotropy of APDs is oppositely signed compared to the main phase or homoepitaxial GaP(100) as sketched in Fig. 5.7. Similar to LEED, RAS measurements laterally average over the microscopic domain structure and the APD content of a GaP/Si(100) sample should effectively reduce the observed RAS signal intensity.

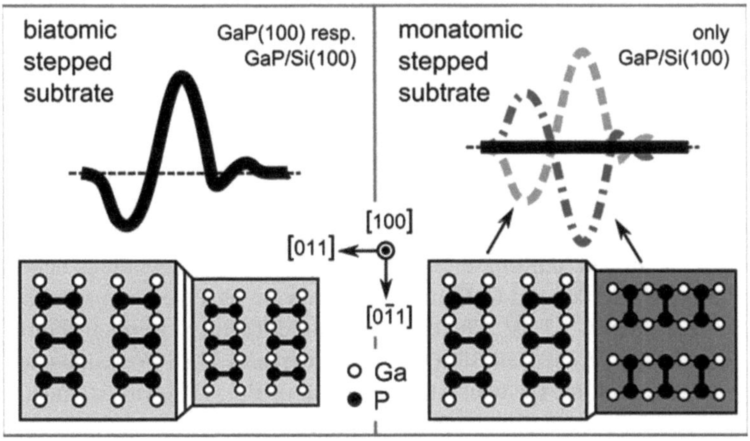

Fig. 5.7 Schematic diagram of P-rich GaP(100) surfaces: the crystal axes define the orientation of the surface reconstruction formed by H-stabilized P-dimers and the sign of the RAS signal. Anti phase disorder causes domains of perpendicular dimer orientation on which the sign of the signal is reversed (*right*). RAS measurement integrates over macroscopic areas of the surface, effectively resulting in a degradation of the intensity corresponding to the amount of APDs on the surface (From [16])

The present study reports on the application of RAS for quantitative in situ analysis of III–V/Si(100) anti-phase disorder and utilizes the pseudomorphic growth of thin heteroepitaxial GaP layers as a model system. The evaluation of the APD content of a distinctly two-domain GaP/Si(100) sample in reference to another heteroepitaxial GaP film of equal thickness leads to promising agreement with ex situ AFM measurements [3]. In contrast, RA spectra of undoubtedly single-domain GaP(100) cannot directly be applied as reference surface, since interference with additional reflections at the GaP/Si(100) heterointerface alters shape and intensity of the sample spectra.

As shown by detailed signal analysis, the major share of the characteristic deviations originates from incorrect normalization of anisotropic surface signals. The determination of the relative reflectance of a sample compared to a P-rich prepared GaP(100) reference surface enables application of corresponding corrections [6] of the RAS data of heteroepitaxy experiments to obtain an accurate APD quantification independent of the GaP film thickness. Extensive benchmarking versus microscopically resolving and statistically evaluated ex situ reference techniques such as AFM and dark-field TEM confirms the reliability and accuracy of the in situ RAS quantification of III–V/Si(100) anti-phase disorder [16]. Using specific film growth conditions for straight propagation of APBs perpendicular to the surface, we simultaneously obtain an indirect in situ quantification of Si(100) surface reconstruction domains [17], since the APD structure of a GaP/Si(100) surface directly reproduces the distribution of surface structure of the Si(100) substrate prior to heteroepitaxy as evidenced by combined AFM and dark-field TEM ex situ measurements. However, bridging the gap between the local information of such microscopically resolving APD characterization techniques on the one hand and the macroscopic information of laterally integrating surface characterization techniques such as in situ RAS and LEED on the other hand usually requires extensive statistics. LEEM dark-field imaging provides an alternative for direct evaluation of III–V/Si(100) anti-phase disorder on a mesoscopic length scale with a non-destructive surface sensitive microscopy technique [15]. Subsequent to their full empirical correction with respect to interference in the normalization and anti-phase disorder, still systematic deviations between the RA spectra GaP/Si(100) and GaP(100) samples were observed, in particular in the spectral range around 3.2 eV. The characteristic differences were attributed to the optical anisotropy of the GaP/Si(100) heterointerfaces [18]. We conducted complex model calculations, for which the approach of Ref. [19] was extended for the consideration of anti-phase disorder, to separate the contribution of the surface and interface dielectric anisotropies (SDA and IDA, respectively) to the RA spectra of our samples.

5.2 Potential for In Situ Detection of Anti-phase Disorder

An in situ technique for reliably determining the single-domain character of large-area surfaces is essential for the application of GaP/Si(100) quasi-substrates. Typical deposition conditions in the MOVPE reactor require optical in situ techniques instead of e.g. probing with electron beams [20]. In situ RAS of GaP(100) and its atomic surface structures was the subject of several recent studies. Different terminations have been identified and were assigned to characteristic RA spectra [11, 12, 21]. The atomic structure of the well-ordered P-rich GaP(100) surface that is essentially formed by buckled P-dimers each stabilized by a single hydrogen atom and that was also observed on P-rich InP(100) [13], is typical for standard MOVPE preparation. Here, RAS was considered to measure the APD density over the entire spot size (about 0.5 cm^2) [3]. According to its definition, RAS signals originating from alternating, vertically aligned domains (e.g. APDs) are reversed. Thus, degradation of the RAS signal intensity occurs corresponding to the relative amount of APDs on the surface of the deposited film. In contrast, on Si(100) substrate prepared preferentially with biatomic steps as intended for the substrate of sample **D** (Fig. 5.10b), the GaP/Si(100) surface produces a far less distorted RAS signal with significantly higher amplitude.

In Fig. 5.9 in situ RA spectra of thin, hetero-epitaxial GaP layers grown on Si(100) are presented. 40 nm thick GaP films are grown on VPE prepared Si(100) substrates intended to exhibit either single-domain or two-domain surface reconstructions, respectively. In Ref. [22] it was described how to prepare preferentially single-domain GaP/Si(100) surfaces. Key issues are the substrate off-orientation, the temperature and the partial pressure of hydrogen during a specific annealing step of the Si-epitaxial buffer layer as well as the specific preparation parameters for the III–V layer. Hence, here two different preparation routes were applied: The first sample (**D**) was prepared with optimized parameters for creating double(D_B)-stepped, single domain GaP/Si(100) surfaces by employing slightly misoriented Si(100) substrates of about 0.1° towards the [011] direction and preparation parameters of Ref. [22]. To prepare the second sample (**S**) Si(100) substrates with a slight misorientation (0.1°) towards the [010] were employed resulting in preferentially two-domain (single-stepped) Si-surfaces [22]. The precursors tertiarybutylphosphine (TBP) and trimethylgallium (TMGa) were used to grow thin, hetero-epitaxial GaP films. An adequate nucleation layer was generated by pulsed, alternating injections of the precursors at 450 °C, while the actual growth of the film was performed at 575 °C. The process was stopped well before reaching the relaxation thickness of the GaP top layer yielding pseudomorphic films of about 40 nm thickness as verified by high-resolution X-ray diffraction. The GaP/Si(100) samples were cooled down in the presence of TBP in order to prepare the P-rich surface of GaP(100) [11]. The whole preparation was monitored by in situ RAS employing a LayTec EpiRAS 200 spectrometer.

In particular, it was applied to control the MOVPE preparation of the well-known P-rich GaP(100) surface for all samples [11, 13]. In order to quantify the amount of residual APBs in the different experiments, at first a standard, P-rich GaP(100) reference layer **R** was grown homoepitaxially on a GaP(100) substrate (Fig. 5.8, spectrum **R**). All RA spectra of Figs. 5.8 and 5.9 exhibited the principle RAS features observed on the homo-epitaxial GaP(100) reference sample **R**. The RA spectrum **D** presented in Figs. 5.8 and 5.9 was taken on GaP/Si(100), when the silicon substrate was preferentially prepared as single-domain surface corresponding to results of Refs. [8, 22]. The RA spectra of both samples (**R** and **D**) were compared in Fig. 5.8 and their amplitudes agree almost perfectly. In particular, the peak intensities are nearly identical suggesting a GaP/Si(100) surface essentially free of APDs (sample **D**). In comparison to the RA spectrum of the single-domain GaP/Si(100) surface (**D**), the corresponding spectrum of the GaP/Si(100) surface, that was intentionally prepared on a two-domain Si-substrate (**S**),

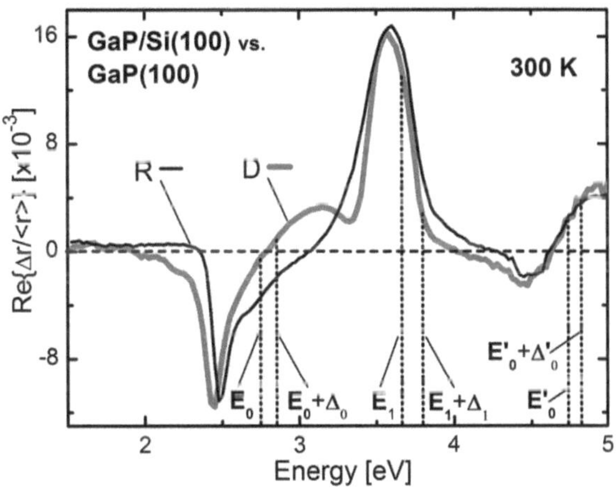

Fig. 5.8 Comparison of the RA spectra of sample *D* and of a homoepitaxially prepared GaP(100) reference *R* (From [3])

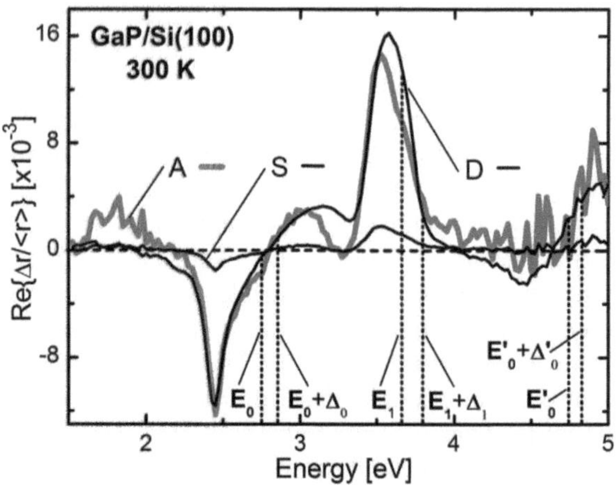

Fig. 5.9 300 K RAS of a single-domain GaP/Si(100) surface (spectrum D), the corresponding GaP/Si(100) surface that was intentionally prepared as two-domain surface (spectrum S) and its eightfold amplification (spectrum A) (From [3])

shows a much lower intensity, as is shown in Fig. 5.9. It verifies that the RAS signal intensity taken on the thin GaP(100) films strongly depended on the Si(100) preparation route and, hence, on the distribution of reconstruction domains on the surface. Due to the superposition of inverted RAS signals originating from different APDs, the observed RAS peak intensities allow for a quantitative analysis of the APD concentration on the GaP(100) surface. For this purpose, Fig. 5.9 also shows another spectrum (**A**), which is an eightfold amplification of spectrum **S**. Except for some minor differences (see below), the amplified spectrum **A** agrees very well with the spectrum of the single-domain surface from sample **D**. Thus, the RAS intensity of spectrum S was only about 1/8 (12.5 %) of spectrum **D** from the single domain GaP/Si(100) surface. This is explained by the difference of the domain distribution on the surface of sample **D**, indicating an APD imbalance and, thus, corresponding to an APD ratio of about 44–56 % on the surface of sample **D**.

Figure 5.8 also displays some differences between the RA spectra of samples **D** and **R**. While spectrum **R** of the homo-epitaxial reference represents a standard RA spectrum of P-rich GaP(100), only originating from the atomic surface reconstruction [13], in experiment **D** and **S** additional contributions to the RAS signal $\Delta r/r$ arose from reflections on the buried GaP/Si(100) hetero-interface. These interface reflections induced interferences of the total reflectivity, which was utilized for the normalization of the RAS signal, and might also be modified by some additional minor optical interface [23] and strain-induced anisotropies [24]. However, beyond the direct band gap of GaP (2.8 eV) these interferences diminish towards higher energies due to absorption, in particular, at critical points in the Brillouin zone like the bulk-induced interband transition E_1, at 3.7 eV, where the strongest RAS peak is induced by a so-called surface-to-bulk transition [25]. In this study, the induced experimental error regarding the RAS peak intensity at around 3.6 eV was measured below 2 %.

Ex situ AFM images of the GaP/Si(100) samples **D** and **S** and the corresponding Si(100) substrate surfaces are shown in Fig. 5.10. Both images exhibit smooth

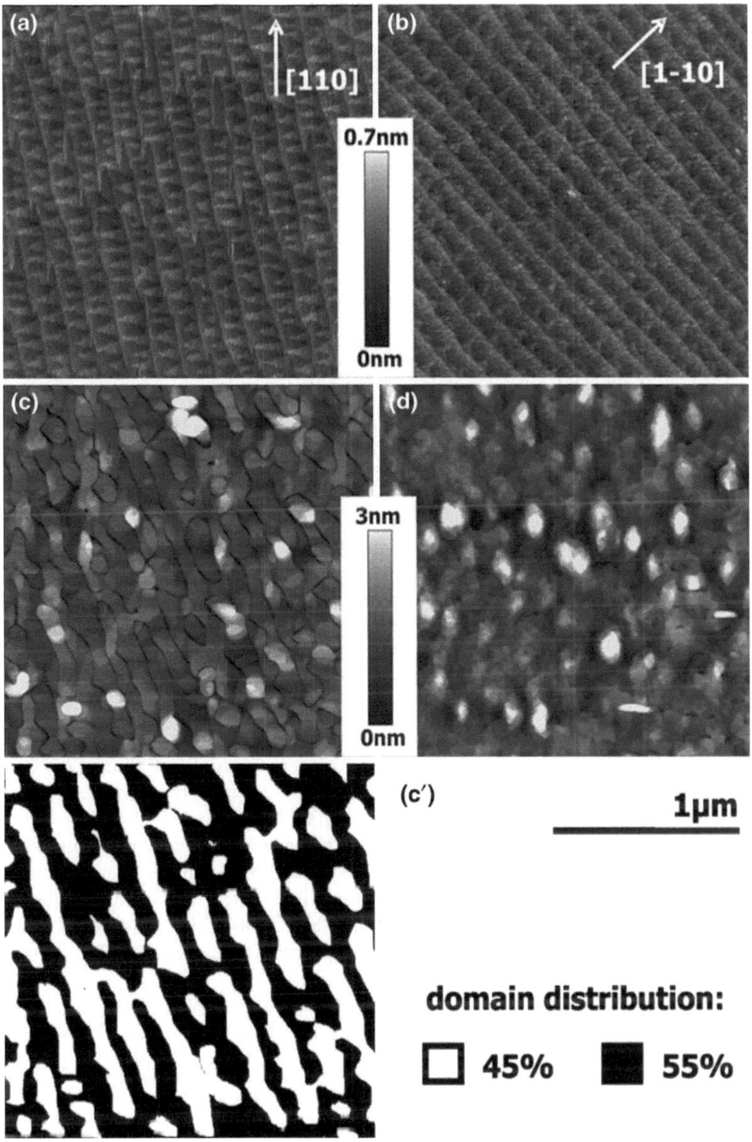

Fig. 5.10 AFM of Si(100) before (**a**, **b**) and after GaP/Si(100) growth (**c**, **d**): in contrast to sample **D** (**d**), the surface of sample **S** (**c**) exhibits a network of anti-phase boundaries after GaP/Si(100) growth. In (**c′**) the two different surface domain-types of sample **S** (**c**) are color coded in *black* and *white* to quantify the surface area covered by each phase (From [3])

surface morphologies. However, the surface of sample S shows a network of black lines, which were identified with APBs [8], while the surface of sample D is free from any APB-related characteristics only exhibiting GaP-bilayer surface structures, clearly supporting the above findings of the RAS studies. In addition, in the

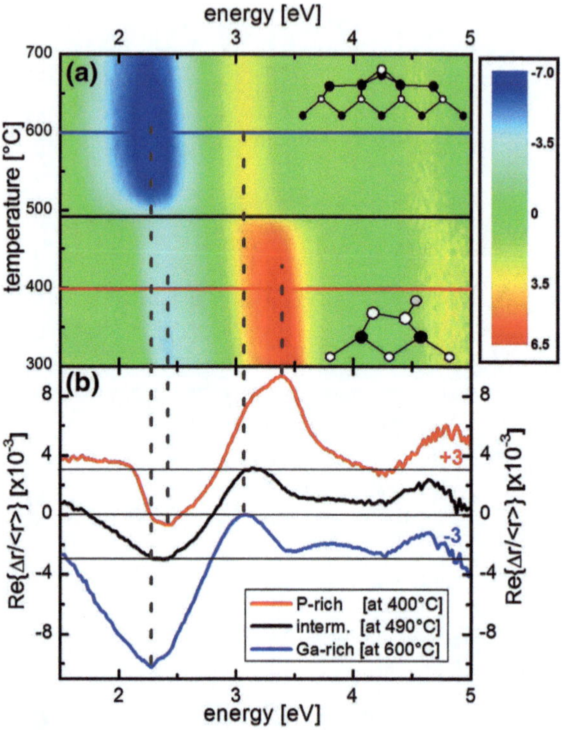

Fig. 5.11 **a** In situ RAS colorplot of the surface of a heteroepitaxial GaP film which was grown on 2° off-orientated Si(100): in direct analogy to Fig. 5.33, spectra were taken continuously while the sample temperature was ramped up from 300 to 700 °C at a rate of 1.5 K/min without TBP supply. Qualitatively, we observed the P-rich surface reconstruction between 300 and 490 °C, the Ga-rich reconstruction beyond 500° C and an intermediate surface state in a narrow range around 490–500 °C as expected in analogy to homoepitaxial GaP(100). **b** Extracted RA spectra which represent the P-rich (*top*), the Ga-rich (*bottom*) and the intermediate surface (*middle*). *Dashed vertical lines* connect the peak positions with the respective features in the colorplot (From [6])

AFM micrograph of Fig. 5.10c′ the surface area covered by one phase was color coded in white to quantify the surface area of the two variants of GaP-APDs on sample S. It revealed a domain ratio of 55–45 % and, hence, perfectly supported the in situ RAS analysis regarding the signal intensities in Fig. 5.8, where a domain ratio of 56–44 % has been determined.

These results clearly underline the chosen approach to quantitatively verify the domain structure of III–V films grown on non-polar substrates for large surface areas, as demonstrated here for the example of GaP/Si(100). For this purpose, optical in situ control with RAS was employed and the linear combination of signals of the different domains was analyzed in terms of the RAS peak intensity. These well established peaks both from experimental as well as theoretical point of view can be regarded as surface sensitive fingerprints and can be employed for benchmarking domain-analysis of surfaces via homo-epitaxial prepared reference samples.

5.3 Quantification of GaP/Si(100) Anti-phase Domains

As the principle of RAS probes the difference in the normal-incidence reflectances R_x and R_y of linearly polarized light directed onto the sample's surface [20, 26] the measurements are well suited to the in situ detection of anti-phase disorder occurring during polar on non-polar heteroepitaxy [3]. According to their orthogonal orientation, the surface area of anti-phase domains (APDs) produces oppositely signed RAS signal contributions $\Delta R = R_x - R_y$ (see Fig. 5.12, compact curve versus dashed curve). Since the measurement integrates macroscopically over the sample's surface, the intensity of the RA spectra also represented an in situ indication of the APD content. However, deducing reliable quantitative information required adequate references and the precise separation of different contributions to the RAS signal. In the case of the pseudomorphic growth of thin GaP films on Si(100) [3, 8, 22] via MOVPE with sufficient P supply, we expected atomically ordered P-rich GaP(100) surfaces after deposition, which should qualify homoepitaxial samples grown under standard conditions as a reference. The results of Chap. 4 justify this straightforward approach, as long as identical surface preparation and measurement conditions for the utilized RA spectra ensured that other sources of signal changes were excluded.

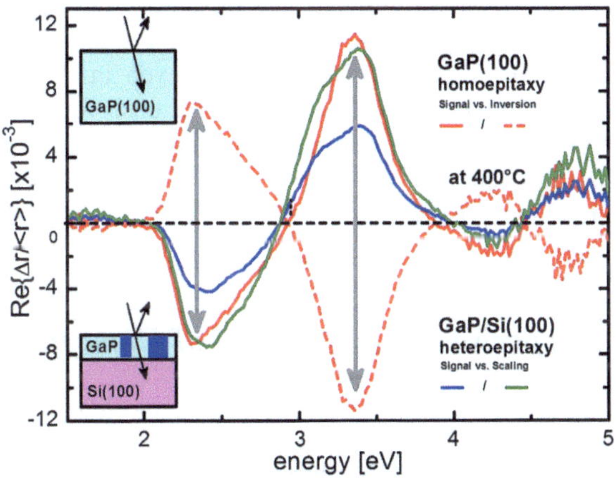

Fig. 5.12 RA spectra extracted from the colorplots presented in Fig. 4.3 (curve with highest peak at 3.4 eV) and Fig. 5.11 (*curve with smallest peaks*) at 400 °C each. The preparation ensured the atomically ordered (2 × 2)/c(4 × 2) reconstruction of both surfaces. Regarding the in situ RAS quantification of APD, the spectrum of the homoepitaxial GaP(100) (*curve with highest peak*) is regarded as reference. Its inversion spectrum (*dashed curve*) arises when the surface structure is turned by 90°. The curve with smallest peaks represents a GaP/Si(100) layer structure containing APD, which locally turn the surface structure by 90°. It was also scaled up to roughly the amplitude and shape of the reference to provide a scaling factor of m = 1.8 for the APD quantification indicating a concentration of about 22 % (From [6])

5.3.1 In Situ Characterization of Heteroepitaxial GaP/Si(100)

For Fig. 5.11, we performed an annealing experiment in analogy to Fig. 4.3 starting from a 40 nm thin GaP film grown on vicinal Si(100), which was about 2° off-oriented towards the [011] direction, to observe the principal surface reconstruction characteristics of an heteroepitaxial sample with in situ RAS. According to the established procedures presented in Sect. 4.2, the P-rich surface was prepared on a heteroepitaxial GaP/Si(100) sample directly after growth. Subsequently, RA spectra were measured continuously, while the temperature was ramped up very slowly from 300 to 700 °C under pure hydrogen carrier gas.

In direct analogy to Fig. 4.3, the results indicated an almost abrupt transition between the P- rich and Ga-rich surface reconstructions at a critical temperature of about 490 °C [27] (Fig. 5.11). Qualitatively, the color-coded RAS data of both experiments [Figs. 4.3 and 5.11 (upper parts)] agree very well, such that we could exclude strain induced modification of the principal surface reconstruction properties. Only a slight difference at the beginning of the experiments (T < 400 °C) resulted from the different starting conditions: Starting with a surface preparation with (Fig. 4.3) or without (Fig. 5.11) excess phosphorus on top, respectively, demonstrated its initial influence on the RAS signal. The selected spectra representing the P-rich, the Ga-rich, and the intermediate states of the GaP/Si(100) surface (Fig. 5.11, lower part) show the basic features as already observed at homoepitaxial GaP(100) (Fig. 4.3). Nevertheless, even in that case, where we prepared single-domain GaP on Si(100) [3], we still observed considerable deviations between the data of hetero- and homoepitaxial samples. These minor, but systematic qualitative discrepancies between selected spectra of Figs. 4.3 and 5.11 will be discussed below (Fig. 5.12).

5.3.2 Direct Comparison to RA Spectra of Homoepitaxial GaP(100)

In spite of the good qualitative agreement of both experiments (Figs. 4.3, 5.11), there was a significant disparity with regard to the scaling of the color-coded representations of the data: the RAS amplitude of the GaP/Si(100) sample (Fig. 5.11) reached only about two-thirds of the amplitude that was associated to the homoepitaxial reference (Fig. 4.3). On the one hand, this basic loss of RAS intensity in the experiment indicated the presence of some anti-phase disorder at the final heteroepitaxial GaP/Si(100) surface [3], since the signal according to its definition ($\Delta R = R_x - R_y$) is reversed on APDs. On the other hand, the observation of a significant RAS intensity in Fig. 5.11 already suggested a preferential domain orientation on the surface, since an equal share of APDs would simply result in the absence of any RAS intensity.

In detail, single atomic layer steps of the Si(100) substrate surface prior to GaP heteroepitaxy initiated the APBs [3, 8], which then confined the APDs. In consequence, the undesired APD content may be reduced either by avoiding APB formation before growth (requiring a bilayer step configuration of the substrate surface) [8] or by eliminating APBs during growth (requiring promotion of pair wise APB self annihilation) [22]. Either way, the GaP/Si(100) surface studied in Fig. 5. 11 already featured a distinct domain imbalance: The effective RAS amplitude of about two-third (Fig. 5. 11) of the reference (Fig. 4.3) suggested roughly one-third of 'inactive' surface area, implying that APDs covered half that share. For the precise evaluation of this quantification scheme, Fig. 5.12 compares the two 400 °C spectra, which were directly extracted from the respective annealing experiments (Figs. 4.3, 5.11).

For the direct comparison of spectra, we selected the atomically ordered P-rich surface reconstructions of the homoepitaxial GaP(100) and heteroepitaxial GaP/Si(100) surfaces, because they combined straightforward MOVPE preparation with an intense characteristic RA spectrum (see Sect. 4.2). Observing the surfaces at T = 400 °C without TBP supply ensured the complete desorption of excess phosphorus and the stability of the well-defined P-rich (2 × 2)/c(4 × 2) surface reconstruction.

Hence, the observed disagreement of the RA spectra in Fig. 5.12 (curve with highest peak versus curve with smallest peaks) was only caused by the impact of the different substrate types used in the experiments. Firstly, different domain orientations produced RA spectra with opposite sign (Fig. 5.12, compare curve with highest peak versus dashed curve). Secondly, the laterally integrated experimental RA spectrum of the heteroepitaxial GaP/Si(100) surface (curve with smallest peaks) reflected its APD content as a superposition of its respective proportional domain areas on the surface. In effect, its intensity appeared reduced compared to the spectrum of the homoepitaxially grown, APD free GaP(100) sample (curve with highest peak), which acted as our reference. Thirdly, the scaling of the heteroepitaxial RA spectrum by a factor of $m = 1.8$ to roughly the intensity of the homoepitaxial one led to a similar shape of the spectra: Thus, the required scaling factor of $m = 1.8$ indicated that, effectively, only a surface area of $a_{eff} = 1/m \approx 56\,\%$ actually contributed to the RAS signal, while APDs covered half of the remaining area then:

$$a_{APD} = \frac{1}{2}\left(1 - a_{eff}\right) = \frac{1}{2}\left(1 - \frac{1}{m}\right) = \frac{m-1}{2m} \approx 22\,\% \tag{5.1}$$

The residual discrepancy between the RA spectra of the heteroepitaxial sample in reference to the homoepitaxial one (compare curves with highest peaks in Fig. 5.12) originated from the differences in the optical structure of the samples (see insets in Fig. 5.12). Aside from the APD content, both surfaces were prepared identically, i.e. (2 × 2)/c(4 × 2) P-rich reconstructed. In contrast to the homoepitaxially grown GaP(100) sample, the GaP/Si(100) sample featured a buried heterointerface. During the RAS measurement, an additional internal reflection of the transmitted light occurred here, enabling interference with the signal reflected at the surface. Experimentally, the APD quantification accuracy increased, when our referencing procedure employed in Ref. [3] was applied, where an adequate heteroepitaxial single domain sample and ex situ AFM analysis were used to benchmark the APD content. A desired in situ process control of the APD content, however, required a less complex routine as well as a reference concept that combines reliability and accessibility. In principle, the straightforward MOVPE preparation of P-rich surfaces on homoepitaxial GaP(100) met these requisites quite well (Sect. 4.2) and in the following a more detailed understanding of the signal deviations is given to assure the APD quantification results and enhance their accuracy.

5.3.3 Impact of the GaP/Si(100) Heterointerface on In Situ Spectra

In Ref. [3] we already determined optical interference of the surface reflection with interfacial reflections as the origin of the RAS deviations, but the microscopic and detailed origin remained unclear. While the standard RA spectrum $(\Delta r/r)_{ref}$ of the P-rich GaP(100) reference *ref* (Fig. 4.3) originated from the atomic surface reconstruction only, additional RAS contributions arose from the buried hetero-interface of the GaP/Si(100) sample *spl* (Fig. 5.11). According to their phase relation, the optical interference in the RA spectra depended on the wavelength λ, on the GaP film thickness *d*, on its refractive index $n = n(\lambda)$, and on the phase shifts associated to the reflections combined in the interference factor $f = f(d, \lambda, n(\lambda))$. In principle, the thin GaP film acted as a Fabry-Pérot type resonator and both constituents of the sample's effective RAS signal Δr and r

$$\left(\frac{\Delta r}{r}\right)_{spl} = \frac{\Delta r_{eff}^{spl}}{\left\langle r_{eff}^{spl}\right\rangle} \tag{5.2}$$

potentially contained significant interference effects. For simplicity of the following expressions, we present a first order approximation of our considerations, where multiple reflections were neglected. The reflections at the surface *surf* and at the interface *int* both contributed to the total reflection of the entire structure:

$$\left\langle r_{eff}^{spl}\right\rangle = \left\langle r_{surf}^{spl}\right\rangle + f \cdot \left\langle r_{int}^{spl}\right\rangle \tag{5.3a}$$

Independently, interference might also occur regarding the reflection difference:

$$\Delta r_{eff}^{spl} = \Delta r_{surf}^{spl} + f \cdot \Delta r_{int}^{spl} \tag{5.3b}$$

Finally, both terms were used for the calculation of the actual RAS signal according to Eq. 5.2. In consequence, the pure presence of an interfacial reflection ($\langle r_{int}^{spl} \rangle \neq 0$) implied significant impact on the RA spectra due to the normalization, even if the interface reflection itself did not contain any anisotropy ($\Delta r_{int}^{spl} = 0$). The incident light was partially reflected at the surface and at the interface. The latter part, of course, also passed twice through the GaP film and its surface. Accordingly, the term $\Delta r_{int}^{spl} = 0$ actually represented a sum of many potential sources of optical anisotropies (as shown in Eq. 5.3c) including: a specific atomically ordered GaP/Si(100) interface formation leading to an interface anisotropy $\Delta r_{int}^{spl}(i)$ [23], a strain induced bulk anisotropy of the GaP film $\Delta r_{int}^{spl}(b)$ [28], an anisotropy caused by space charge regions [29] due to the polar on non-polar growth $\Delta r_{int}^{spl}(c)$ [2], mesoscopically ordered anisotropic structures (step edges, APB, twins etc.) $\Delta r_{int}^{spl}(m)$ [30], and others:

$$\Delta r_{int}^{spl} = \sum_j \Delta r_{int}^{spl}(j) \tag{5.3c}$$

The diversity of these potential contributions to the RA spectra of heteroepitaxial GaP/Si(100) films impeded detailed predictions, of how the APD quantification was affected in comparison to the homoepitaxial RAS signal. An experimental approach to separate different origins of the total anisotropy from each other could simplify the identification of specific contributions. According to their practical relevance, finally, a model containing adequate corrections might enhance the quantification accuracy. So, our further consideration essentially required a precise assignment, to which extent the observed deviations from the homoepitaxial anisotropy were caused either by the normalization of the signal (due to interferences regarding $\langle r_{eff}^{spl} \rangle$) or by actual modifications of the optical anisotropy (contributing in proper phase relation to Δr_{eff}^{spl}), e.g. by an interface anisotropy.

5.3.4 Interference Induced by the GaP/Si(100) Heterointerface

The measured absolute reflection spectra $\langle r_{eff}^{spl} \rangle$ principally contained two different kinds of wavelength dependencies: firstly, the pure emission spectrum $E^l(\lambda)$ of the utilized light source l (in our experiment a Xe lamp) and secondly the specific, material dependent reflectance spectrum $R_{abs}^{spl}(\lambda)$. Due to the convolution of these spectra $\langle r_{eff}^{spl} \rangle = R_{abs}^{spl}(\lambda) \cdot E^l(\lambda)$, continuous temporal changes corresponding to the aging of the lamp $E^l(\lambda, t)$ also contributed to the data. Therefore, we measured both, the sample *spl* and the reference *ref* within a reasonable time period t_i to derive a precise relative reflectance $R_{rel}^{spl}(\lambda)$:

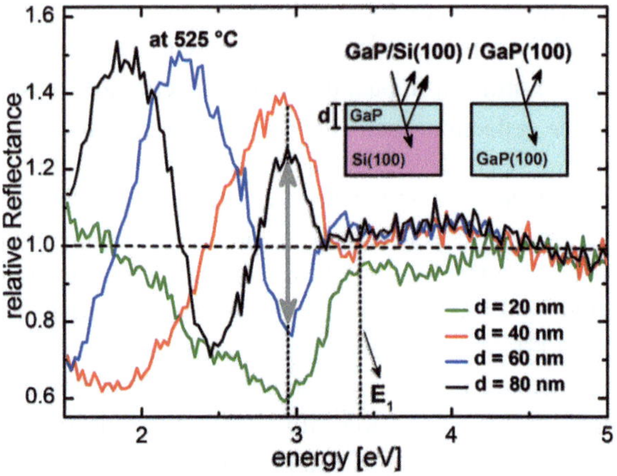

Fig. 5.13 Relative reflectance spectra $R_{\rm rel}^{\rm spl}(\lambda)$ (see text) calculated for the heteroepitaxial growth of a GaP film on a Si(100) substrate in relation to a homoepitaxial GaP(100) reference with identical surface preparation conditions. The individual data sets for different GaP thickness values were taken in growth interruptions with continued TBP stabilization. Due to the optical path difference between the light reflected on the surface and the light reflected at the GaP/Si(100) interface Fabry-Perót interference occurred, which vanished at higher photon energies due to increasing absorption by the GaP film, in particular beyond the E_1 interband transition. Variation of only 20 nm film thickness turns a maximum into a minimum and vice versa at around 3 eV (From [6])

$$R_{rel}^{spl}(\lambda) = \frac{R_{abs}^{spl}(\lambda)}{R_{abs}^{ref}(\lambda)} = \frac{R_{abs}^{spl}(\lambda)\cdot E^{l}(\lambda, t \in t_i)}{R_{abs}^{ref}(\lambda)\cdot E^{l}(\lambda, t \in t_i)} = \frac{\left\langle r_{eff}^{spl}\right\rangle}{\left\langle r_{eff}^{spl}\right\rangle} \tag{5.4}$$

Equation 5.4 shows that the knowledge of the absolute reflectance of the reference $R_{abs}^{ref}(\lambda)$ in principle allows the calculation of values for sample $R_{abs}^{spl}(\lambda)$. But in practice $R_{rel}^{spl}(\lambda)$ already provided direct access to the thin film interferences (as shown in Fig. 5.13), since the surface structures of the homoepitaxial GaP(100) and the heteroepitaxial GaP/Si(100) were identical, which was ensured by well-defined preparation procedures (Sect. 4.2).

In both cases, a specific share of the probe light was reflected at the atomically ordered P-rich surface reconstruction of the epitaxial GaP(100) layers. In addition, only for the heteroepitaxial samples, a part of the initially transmitted light was reflected at the buried GaP/Si(100) interface and eventually reached the detector. The phase relation between both contributions explicitly varied with the considered wavelength of light λ and the actual thickness d of the GaP film giving rise to so-called Fabry-Pérot oscillations within the relative reflectance spectra of the respective samples $R_{rel}^{spl}(\lambda)$ as shown in Fig. 5.13.

The results in Fig. 5.13 proved the presence of significant interferences in the reflectance of GaP/Si(100) samples relative to a homoepitaxial reference and gave a first insight of the magnitude of the modification of the respective RA spectra via the normalization by the reflected light $\langle r_{eff}^{spl} \rangle$. The high refractive index of GaP [31] led to significant oscillations even for moderate film thicknesses. According to the optical path difference Δ, the first destructive interference ($\Delta = \lambda/2$) occurred after about 20 nm GaP growth at photon energies of about 3.0 eV and shifted to longer wavelength during further deposition. At the same time, constructive interferences ($\Delta = n \cdot \lambda$) and destructive interferences ($\Delta = (2n-1) \cdot \lambda/2$) of higher orders occurred. In effect, the interference conditions at the fixed position of 3.0 eV alternated in steps of $d = 20$ nm film thickness (Fig. 5.13), which is often utilized for standard in situ growth analysis with a monochromatic light source [32]. At around 525 °C, the E_0 interband bulk transition of GaP is located in the energetic range of 2.7 eV [33]. Beyond this position, the absorption coefficient of GaP starts to rise steeply [34]. This led to a drastically diminished intensity of the interface reflection beyond 3.2 eV photon energy, where the interference effects vanished and the reflectance of the samples equaled that of the homoepitaxial GaP(100) reference.

5.3.5 Interference Corrected Normalization of In Situ RAS Signals

These findings implied several severe consequences for the in situ quantification of GaP/Si(100) APDs utilizing the RAS technique. In the optical range between about 1.5 and 3.2 eV photon energy, the reflectance of heteroepitaxial samples changed significantly depending on wavelength and film thickness. The relative reflectance $R_{rel}^{spl}(\lambda)$ in comparison to our homoepitaxial reference (Fig. 5.13) roughly ranged from 0.6 to 1.5 just due to simple Fabry-Pérot type interferences. As a direct result, the normalization eventually modified the RAS intensity of arbitrary GaP/Si(100) samples by the inverse of the relative reflectance. Explicitly, both major RAS peaks of the P-rich GaP(100) surface were also affected significantly (Figs. 5.12, 5.14). For a worst case scenario (not shown here), we calculated an impact on the APD quantification via the comparison to the homoepitaxial reference of up to ± 20 % APD content just due to inadequate normalization (still neglecting additional optical anisotropies in the heteroepitaxial samples). In Ref. [3], we applied a sophisticated, indirect referencing procedure, when comparing an especially prepared single domain GaP/Si(100) film with another GaP/Si(100) film of identical thickness but with an unknown APD content. This experimental approach lowered the error limits significantly, but required a precise control over the preparation and characterization procedure including the identical film thickness of two GaP films, one of them completely APD-free, as well as extensive data analysis, and the additional ex situ AFM evaluation. In contrast, the feasibility of an alternative, straightforward APD quantification procedure with high accuracy depended on the reliable separation of the Fabry-Pérot oscillations. An appropriate correction of the

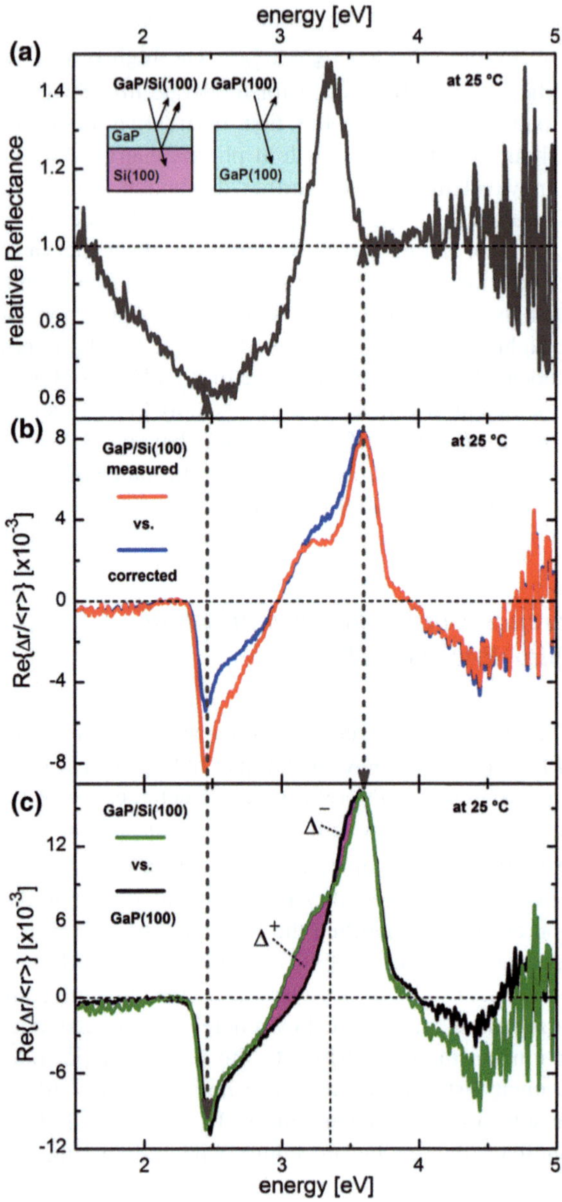

Fig. 5.14 Measurement of an about 30 nm thin GaP film deposited on a Si(100) substrate in comparison to a homoepitaxially grown GaP(100) sample, all conducted at room temperature (25 °C). **a** Relative reflectance $R_{\mathrm{rel}}^{\mathrm{spl}}(\lambda)$ (see text) that was used for correction of the Fabry-Pérot oscillations in the measured data of the heteroepitaxial sample reflection. **b** Comparison of the originally measured RA spectrum (*curve with highest peak* at 2.5 eV) and the corrected RA spectrum (*curve with smaller peak*) according to equation 5.4. **c** Comparison of the corrected RA spectrum that was scaled up by a factor of m = 1.94 to fit the reference RA spectrum taken from an APD free homoepitaxially GaP(100) sample (From [6])

RA spectra prior to a comparison with the homoepitaxial reference also promised qualified indications, whether any significant optical anisotropy regarding either the transmission through the GaP film or the reflection at the buried heterointerface actually contributed to the signal intensity as well.

The deconvolution of the Fabry-Pérot oscillations required the definition of an adequate correction function $F_{cor}(\lambda)$, which reproduced the interference effects. In the given case of identical surfaces for sample and reference, the determination of the relative reflectance like in Fig. 5.13 perfectly fulfilled this requisite [6]. Standard optical software allowed to obtain this correction function with the knowledge of two sample parameters (temperature T and GaP film thickness d). Besides that, the correction function $F_{cor}^{spl}(\lambda) = F_{cor}^{spl}(\lambda, T, d)$ then still depended on the optical data for GaP and Si at the relevant temperatures and, thus, on their availability. Instead, we decided to calculate $F_{cor}^{spl}(\lambda)$ directly from the experimental data according to Eq. 5.4. This approach did not require any additional measurements, since the acquisition of appropriate experimental data (see above) was already part of the measurement. Hence, our route of evaluation avoided the explicit determination of absolute reflectance spectra $R_{abs}^{spl}(\lambda)$ and $R_{abs}^{ref}(\lambda)$. We obtained a straightforward calculation of the Fabry-Pérot corrected RA spectra based on Eqs. 5.2 and 5.4:

$$\left(\frac{\Delta r}{r}\right)_{spl} \cdot F_{cor}^{spl} = \frac{\Delta r_{eff}^{spl}}{\left\langle r_{eff}^{spl} \right\rangle} \cdot \frac{\left\langle r_{eff}^{spl} \right\rangle}{\left\langle r_{eff}^{ref} \right\rangle} = \frac{\Delta r_{eff}^{spl}}{\left\langle r_{eff}^{ref} \right\rangle} \tag{5.5}$$

The advantages of the approach included a reduction of the necessary experimental data input and of the associated error sources. Especially, the calculation of the correction function from the identical experiments led to a principal independence from the GaP film thickness, such that additional in or ex situ measurements are not necessary. The Fabry-Pérot corrected term $\Delta r_{eff}^{spl} / \langle r_{eff}^{ref} \rangle$ actually represents an 'artificial' RA spectrum, whose constituents derived from different measurements conducted at different samples. As desired, only Fabry-Pérot oscillations associated to normalization were deconvoluted from the measured data of the heteroepitaxial samples, but the result still contained all potential sources of additional anisotropy like interface anisotropy. It is also important to note that any additional anisotropy would contribute to the measured RAS signal in a proper phase relation with respect to the surface anisotropy (see below). So far, we obtained a systematic approach of interference compensation with general applicability and explicitly independent of the film thickness.

Figure 5.14 clearly shows the improvement of the in situ RAS quantification of APD by the correction of the Fabry-Pérot interferences following the procedure described above. Here, the RA spectra of an approximately 30 nm thin GaP film deposited on a Si(100) substrate ('sample', Fig. 5.14b, curve with highest peak at 2.5 eV) and homoepitaxial GaP(100) ('reference', Fig. 5.14c, curve with highest peak at 2.5 eV) were measured at ambient temperature (25 °C) to achieve enhanced spectral resolution in comparison to measurement at higher temperatures. Figure 5.14a represents a calculation of the sample's relative reflectance following Eq. 5.4. Obviously, the destructive interference of surface and interface reflections around 2.6 eV (Fig. 5.14a) influenced the normalization of the low energy RAS peak at about 2.4 eV (Fig. 5.14b). In consequence, a strong, artificial amplification of this RAS feature which is only induced by electronic transitions in the surface Brillouin zone, occurred, which directly affected the APD quantification procedure. Meanwhile, the sharper constructive interference at around 3.3 eV (Fig. 5.14a) contributed only at photon energies well below the maximum of the high energy RAS peak at around 3.6 eV. Hence, at the position of the peak which correlates to the energetic position of the bulk critical point E_1 transition, the relative reflectance was close to unity. Thus, the maximum amplitude of the high energy RAS peak was not significantly affected by the interference phenomena (Fig. 5.14a). However, the consideration was constricted to the individual sample, since relatively small variations in the film thickness caused potentially decisive shifts of the interferences (see Fig. 5.13).

According to Eq. 5.5, we carried out the correction of the RA spectrum via the deconvolution of the relative reflectance (Fig. 5.14a). The changes in the shape of the spectrum are depicted in Fig. 5.14b, where the correction result (corrected curve) was compared with the uncorrected RA spectrum as measured (measured curve). Figure 5.14c shows the resulting improvements of the comparability between the RA spectra from the heteroepitaxial GaP/Si(100) sample and the homoepitaxial GaP(100) reference. In analogy to Fig. 5.12, in Fig. 5.14c, the corrected GaP/Si(100) RA spectrum was scaled up to the intensity to the reference spectrum, resulting in an amplification factor of $m = 1.96$. According to the evaluation of Fig. 5.12 and Eq. 5.1, we determined an anti-phase domain content a_{APD} of about 24.5 % on the surface of this sample (Fig. 5.14). The scaling tolerance when matching the corrected RA spectrum with that of the reference equaled an APD quantification error below ±0.8 % (compared to about ±12 % before Fabry-Pérot correction).

Besides variations due to statistical noise during measurement, the potential presence of additional, interface reflection related optical anisotropy according to Eqs. (5.3b, c) still remained a conceivable systematic source of the observed quantification error. Insight into the spatial distribution of APDs and into variations of their concentration over the sample's surface requires investigations with independent methods providing microscopic resolution such as scanning probe techniques such as AFM (Sect. 5.4) and STM) or electron microscopy with specific contrast mechanisms such as TEM (Sect. 5.5.5) and LEEM (Sect. 5.6).

5.3.6 Evaluation of the Empirical Correction Scheme

So far, our analysis of the measured in situ RA spectra of heteroepitaxial GaP/Si(100) samples was focused on the evaluation of (1) the APD concentration and (2) the spectral deformation by interferences contained in the normalization. In reference to the characteristic spectrum of the well-defined P-rich GaP(100) reference, we found adequate compensation procedures for the deconvolution of these specific influence factors:

1. The concept of APD quantification via scaling factors (Eq. 5.1) based on the preparation of identical surface structures, only differing in the content of APDs. When measured under consistent conditions, the concept corresponds to a relation of the surface anisotropies measured at sample and reference given by:

$$\Delta r_{surf}^{spl} = (1 - 2 \cdot a_{APD}) \cdot \Delta r_{surf}^{ref} \tag{5.6}$$

2. The well-defined relative reflectance (Eq. 5.4) applied as correction function for the normalization induced Fabry-Pérot oscillations (Eq. 5.5). At the homoepitaxial GaP(100) (reference) sample, obviously only reflection on the surface occurred, as there was no buried interface. Of course, due to identical preparation, the surface reflection of the heteroepitaxial GaP/Si(100) sample (without interface contributions) should be equal:

$$\left\langle r_{eff}^{ref} \right\rangle = \left\langle r_{surf}^{ref} \right\rangle = \left\langle r_{surf}^{spl} \right\rangle \tag{5.7}$$

Utilizing Eq. 5.3b, we combined the results to obtain a transformation formula between the RA spectra of sample and reference, applicable for arbitrary sample structures, which consist of a thin polar film heteroepitaxially grown on a non polar substrate. Assuming suitable measurement conditions (see above) and an adequate choice of reference (requiring the preparation of an identical surface reconstruction) we obtain:

$$\left(\frac{\Delta r}{r}\right)_{spl} \cdot F_{cor} = (1 - 2 \cdot a_{APD}) \cdot \left(\frac{\Delta r}{r}\right)_{ref} + f \cdot \left(\frac{\Delta r_{int}^{spl}}{r_{surf}^{ref}}\right)_{\text{int}} \tag{5.8}$$

Besides the experimental spectra and the correction function, which can be derived from the measurements, Eq. 5.8 consisted of just two variable parameters: the actual concentration of APDs at the surface of the heteroepitaxial GaP film a_{APD} and the complex term $f \cdot \left(\Delta r_{int}^{spl} / r_{surf}^{ref}\right)_{int}$ which expresses the contribution of potential optical anisotropy arising due to the interfacial reflection. Corresponding to Eq. 5.8, the presence of systematic deviations between the reference measurement and the corrected and scaled RA spectrum could prove the existence of an anisotropy related to the interface reflection.

5.3.7 Indications for Optical Interface Anisotropy

The comparison in Fig. 5.14c actually formed an instructive example for our data analysis. Except for some deviations, the shape of the corrected and scaled GaP/

Si(100) RA spectrum [Fig. 5.14c, GaP/Si(100)] agreed very well with that of the homoepitaxial reference [Fig. 5.14c, GaP(100)], which also underlined the precise determination of the scaling factor. However, within the region between about 2.9 and 3.5 eV, Fig. 5.14c also displays a systematic deviation (area between spectra) between correction result and reference: First, the spectral difference (area between spectra) begins positive (in the range below 3.3 eV, marked Δ^{+}) and then turns negative (beyond 3.3 eV, marked Δ^{-}). Although the systematic difference indicated the presence of some additional optical anisotropy in heteroepitaxial GaP/Si(100), it only provided indirect access to the origin of this RAS contribution. First of all, the optical anisotropies of the surface and the interface reflections also interfered during our measurement. Accordingly, the development of the phase relation implied modulations of the intensity and of the sign of the spectral difference. Regarding the deviation's change of sign at about 3.3 eV, either a characteristic feature or just a change in the interference conditions could have occurred at this energetic position. Beyond that, the distinct spectral difference (Fig. 5.14c) was arbitrarily normalized to the reflectance of a P-rich GaP(100) surface and implicitly scaled according to the APD content of the film's surface. Hence, the systematic deviation after the corrected normalization strongly indicated some optical anisotropy associated with the interface reflection, but provided only limited, indirect access to the actual structure of this RAS contribution so far.

Potential sources of interface anisotropy [23] could be a specific interface formation and related electronic transitions. Nevertheless, well-known assumptions of interface charges arising due to polar on non-polar epitaxy [2] could lead to potentially anisotropic electro-optical phenomena [29]. Since the lattice parameters of Si and GaP are slightly different and still change with temperature, also strain induced optical anisotropy of the pseudomorphic GaP layer (like birefringence [28]) has to be considered. Moreover, anisotropic distribution of mesoscopic structures [30] including interfacial steps, and anti-phase boundaries could also affect the RAS signal. Appropriate variation of parameters such as the GaP film thickness, the substrate orientation, and the methodical variation of the nucleation procedure in combination with an advanced analysis of the experimental data in the course of our current work are supposed to clarify this open question, i.e. the origin of the interface related anisotropy.

5.4 In Situ Verification of Single-Domain Polar on Non-polar Epitaxy

In the following, we will compare results on the APD content of GaP layers grown on Si(100) derived from ex situ AFM surface analysis to the results derived from in situ RAS investigation. In order to visualize the APDs via AFM, the GaP/Si(100) sample was annealed after growth. Since a preferential desorption of GaP occurs in the range of APBs due to the weaker bonding and higher strain, annealing results in trenches visible as height contrast in AFM images [8] in

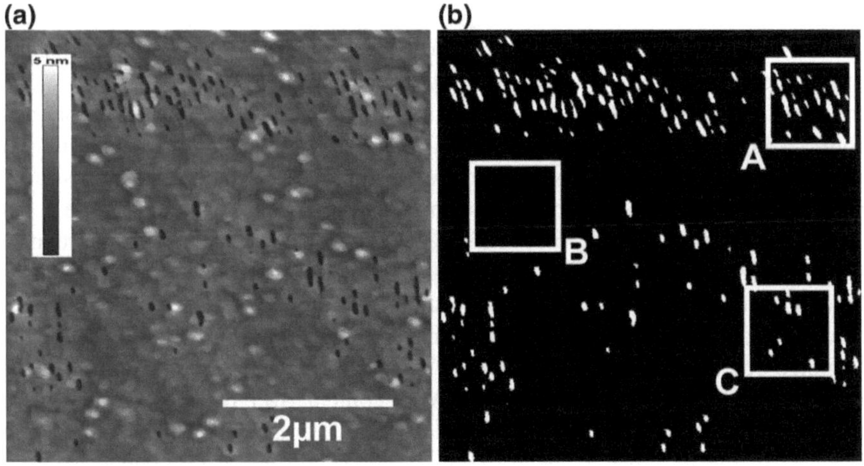

Fig. 5.15 AFM characterization of the GaP/Si(100) surface after specific annealing: the AFM image (**a**) identifies APDs by induced height contrast. The transfer to color-coded image (**b**) highlights the APD content determined to be 3.1 %. *The marked regions* are discussed in the main text (From [16])

analogy to anisotropic etching [7]. A comparison with cross-sectional TEM investigations was used to prove that the position of the dips in the GaP surface indeed corresponded with the location of an APB intersecting the GaP surface [8]. Color-coding of the APDs in the AFM images according to their boundaries allowed direct quantification of the local APD content within the respective image (Fig. 5.15).

Extensive AFM characterization showed that the APD content varied with the position on the sample's surface [16]. Even within certain, sufficiently large images such as Fig. 5.15, we observed significant variations of the local APD concentration. While some areas of the surface appeared completely APD free (Fig. 5.15, pos. B), other spots displayed rather high concentrations of APDs (Fig. 5.15, pos. A). Only few, scattered APDs covered major parts of the measured AFM images (e.g. Fig. 5.15, pos. C), which reflects the average state and the average APD content of the sample discussed here. The observed inhomogeneity of the APD distribution on the surface of this specific sample was a result of a Si buffer preparation, which was intentionally rough. This resulted in an APD structure consisting of differently large APDs during the overgrowth with GaP—depending on the Si surface step distance as well as on the angle and the direction of the local substrate misorientation. As the growth conditions were chosen so that the APDs self-annihilate on {111} or higher index planes, differently large APDs result in different layer thicknesses of complete self-annihilations. Hence, after the 40 nm GaP layer thickness chosen, only those APDs self annihilated and accordingly did not intersect the GaP surface, which had a small enough diameter at their base.

We evaluated several AFM images regarding their APD content to collect sufficient statistics of the sample in order to overcome the local variations of the sample (see Fig. 5.15). The analysis led to a mean value of 3.0 % of APDs and a standard deviation of 1.8 %. The calculation based on the APD coverage obtained from 5 AFM images (each in the size of 5 × 5 μm^2), which were taken at different locations of the surface. The same sample was previously characterized in situ by RAS, which addresses larger macroscopic sample spots (see below). The local variations observed in Fig. 5.15 enforced extensive AFM analysis in order to obtain a mean value for the APD content with adequate statistics, in particular as the choice of growth conditions in this study intentionally yielded an inhomogeneous surface structure. While the implicit lateral integration of RAS was in the cm-range due to the typical probe size of RAS, AFM probes areas within the μm-scale. So, the lateral variations of the APD content over the surface area were rather prominent in AFM images, but they were averaged over the macroscopic spot size in the in situ RA spectra (Fig. 5.16). Therefore, RAS yielded a rather significant value for the sample's APD content.

Figure 5.16 compares the in situ RA spectra of the GaP/Si(100) sample which was subsequently characterized by ex situ AFM (Fig. 5.15) and a APD-free, homoepitaxial GaP(100) reference measured at ambient temperature. According to [6], the preparation procedure strongly affects the resulting surface structure. Thus, identical MOVPE procedures were applied to assure identical atomic surface reconstructions. In absence of any bulk anisotropy, the reference spectrum originates only from the atomic configuration of the surface [13]. However, the buried GaP/Si(100) heterointerface of the sample gives rise to additional RAS contributions. Here, we focus on the effects of anti phase disorder and reflections at interior interfaces. According to Eq. 5.2, anti phase disorder simply reduces the intensity of the peaks assigned to a specific surface reconstruction.

Reflections at interior interfaces modify both the anisotropy $\Delta r = r_x - r_y$ and the total reflection $r = \frac{1}{2}(r_x + r_y)$ in a quite complex manner. The rather good

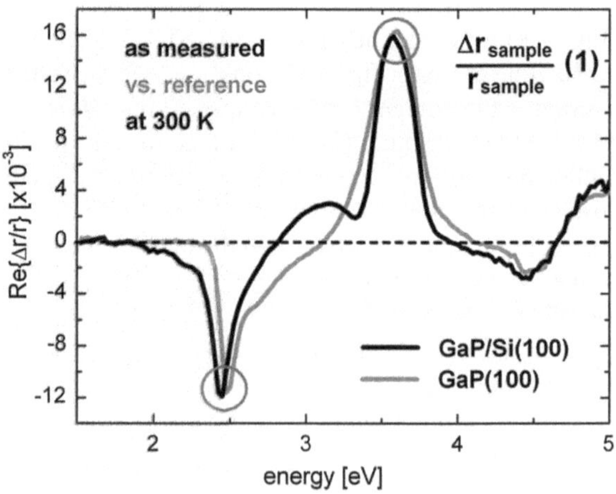

Fig. 5.16 Comparison of the RA spectra measured at an almost single domain heteroepitaxial GaP/Si(100) sample (*black curve*) and at a homoepitaxially prepared GaP(100) reference (*grey*). A comparison of the peak heights indicated low anti phase disorder. Differences of the peak intensities indicated an uncertainty of the APD content in the order of a small percentage (From [16])

agreement of peak intensities in Fig. 5.16 indicated a nearly APD-free GaP/Si(100) film. However, slight deviations between the signals (black and grey curves of Fig. 5.16) still permitted a small APD content of the sample's surface in the order of a few percent. Linear scaling to the intensity of the single-domain reference could not overcome this uncertainty, since the deviations of the two peaks in Fig. 5.16 were of opposite sign. Moreover, ambiguity regarding the actual APD content of the sample arose from interferences with the reflection at the buried GaP/Si(100) interface (Fig. 5.17), so far neglected in the evaluation.

To get an impression of the influence of the internal reflection, we calculated the relative reflectance according to Eq. 5.3a shown in Fig. 5.17. Assuming no influence of the internal reflection, this ratio should be unity over the whole spectrum. In the energy range above 4 eV this is approximately true due to poor transmission through the GaP film. Since the absorption coefficient for lower photon energies is smaller, a significant fraction of the incident light was reflected at the interface and gave rise to interference within the RAS measurement. So, the relative reflectance was unequal to one but varied approximately between 0.6 and 1.6. Due to the normalization of the RA spectra via the total reflection (Eq. 5.1), corresponding intensity modulations affected the sample spectrum directly.

Comparing Figs. 5.16 and 5.17, the relative reflectance was below unity at both peak positions, leading to a slightly underestimated of the APD content by direct scaling of the spectrum prior to the interference correction shown in Fig. 5.18.

We corrected the interference effects by convoluting R_{rel} to the measured RA spectrum [6] (Fig. 5.18). The "artificial" spectrum $\Delta r_{sample}/r_{ref}$ then exhibited

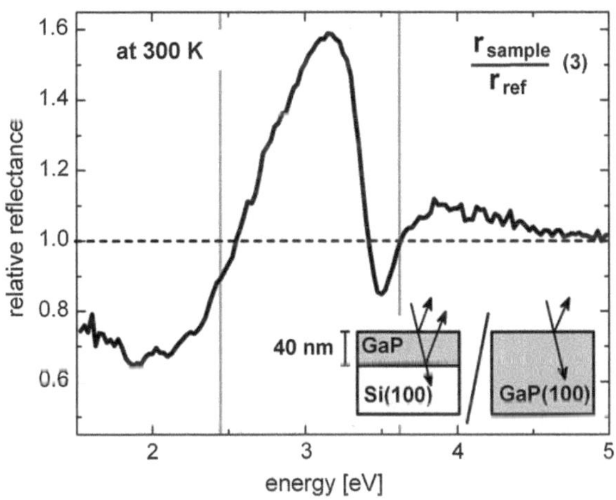

Fig. 5.17 Relative reflectance of the heteroepitaxial GaP/Si(100) sample shown in Fig. 5.16 in relation to the homoepitaxial reference calculated directly from the experimental data according to Eq. 5.9. The deviations from unity indicate the interference with the interfacial reflection and alter the experimental RA spectra via the normalization of the signal. *The gray vertical lines* indicate the energetic peak positions of the measured GaP/Si(100) RA spectrum shown in Fig. 5.16 (From [16])

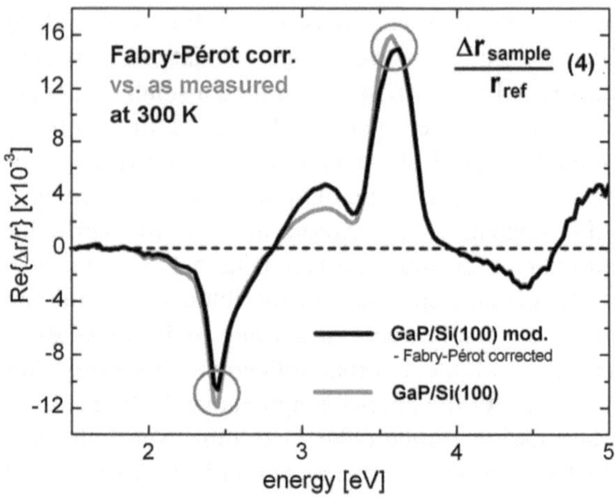

Fig. 5.18 Comparison of the measured RA spectrum from the heteroepitaxial GaP/Si(100) sample shown in Fig. 5.16 with the "artificial" spectrum (see text) according to Eq. 5.4, i.e. the convolution of the measured spectrum with the relative reflectance. Both peaks slightly decrease in intensity, the negative peak more than the positive one (From [16])

slightly less intensity as the reference at both peak positions indicating that the sample in fact was not completely APD free. Linear scaling of the corrected spectrum by a factor $m = 1.09$ resulted in accurate matching peak intensities (Fig. 5.19), which corresponds to an area $a_{eff} = 91.7\,\%$ effectively contributing to the RAS signal. According to Eq. 5.2, we calculated the fraction of the sample's surface covered by APDs to be $a_{APD} = 4.1 \pm 0.8\,\%$. Within the error bars associated with both techniques, this value is in excellent agreement to the APD concentration of $(3.0 \pm 1.8)\,\%$ we obtained by extensive ex situ AFM investigations.

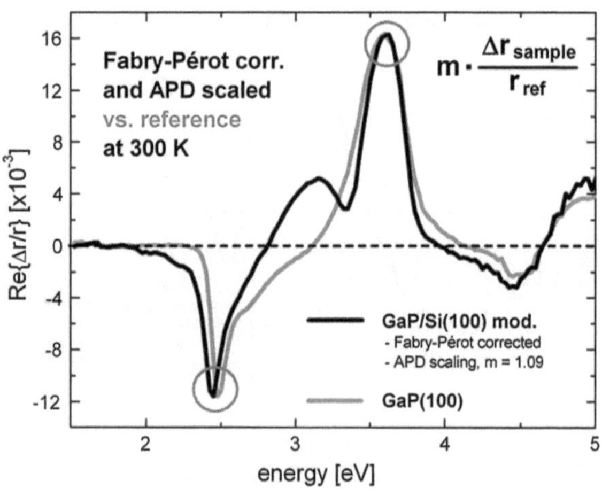

Fig. 5.19 Complete correction of the RA spectrum from the heteroepitaxial GaP/Si(100) sample shown in Fig. 5.16. The convolution of the relative reflectance to the measured RA spectrum corrected the interference influence on the normalization. Afterwards, linear scaling of the calculated spectrum compensated signal reduction caused by APDs. The peaks intensities now match more accurately (From [16])

Besides the improved agreement of the peak intensities, correction and scaling explicitly enlarged the deviation between sample and reference spectrum in the energetic range around 3.2 eV (Fig. 5.19). The applied interference corrections only affected the normalization of the RA spectrum, without any modification of the reflection difference. Thus, any interface anisotropies potentially contained in the interfacial reflection [19, 23] still fully accounted to the corrected spectrum (black curve of Fig. 5.19). Measurements at different GaP film thicknesses (not shown here) indicated a systematical optical anisotropy originating from the GaP/Si(100) heterointerface. Here, further studies combined with thorough optical simulations are required for a clear analysis of this interface anisotropy.

5.5 Indirect In Situ Characterization of Si(100) Surface Structures

In this section, we present a concept for the indirect in situ characterization of Si(100) substrates regarding the concentration of (2×1) and (1×2) surface reconstruction domains applying the RAS quantification of APDs on subsequently grown III–V layers. In contrast to the technological concept to enhance the self-annihilation of APBs [8], their strictly straight propagation in growth direction conserved the atomic surface structure of the substrate, which was transferred into the lateral distribution of the anti-phase disorder in the III–V film. By the example of pseudomorphic GaP growth, we show laterally integrating in situ RAS results for differently prepared Si(100) substrates in comparison to microscopically resolved AFM and TEM measurements regarding the Si(100) surface and the deposited GaP film, respectively.

5.5.1 Surface Structure of the Si(100) Substrates

Since double-layer steps on Si(100) substrates prevent the formation of anti-phase disorder in subsequent III–V heteroepitaxy (Sect. 3.7), preparation routines in ultra-high vacuum are extremely well-studied and involve vicinal Si(100). Although the influence of atomic hydrogen adsorption is also considered in detail [35], the use of hydrogen as typical carrier gas in VPE environments leads to unexpected results: Kunert et al. [22] report a VPE preparation route inducing a preference for double-layer steps for only by 0.1° misoriented Si(100) if tilted in a specific direction. Based on their routine, we obtained different atomic surface structures as shown in Fig. 5.20 just by the choice of Si(100) substrates which essentially only differed in the orientation of the slight off-cut angle.

The AFM images in Fig. 5.20 show the atomic surface structure of Si(100) substrates after the VPE-preparation which usually preceded GaP deposition

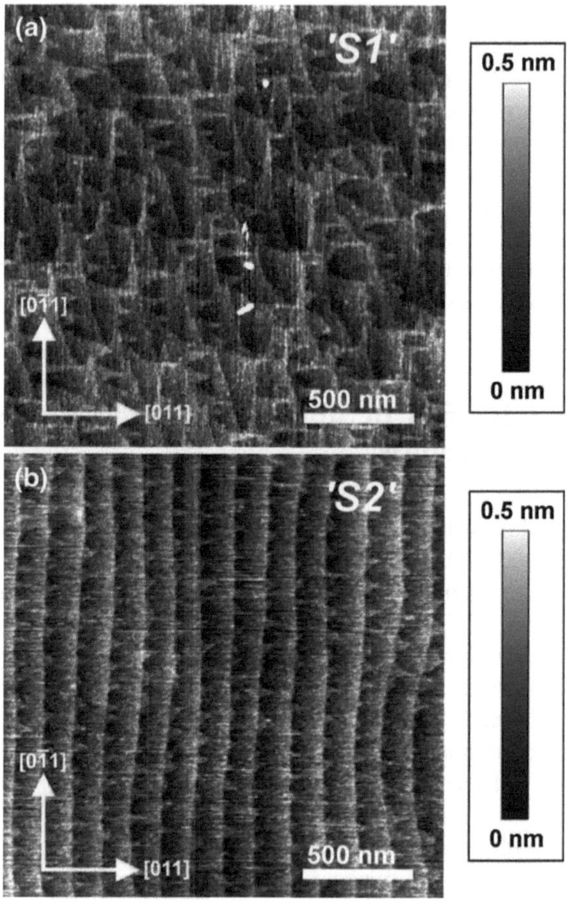

Fig. 5.20 Ex situ AFM characterization of Si(100) surfaces after identical VPE preparation processes applied to about 0.1° misoriented substrates with projection of the off-cut pointing either about 30° off from (type 'S1') and almost exactly in (type 'S2') the [011] direction, respectively: while the first configuration (**a**) led a two-domain surface with complex structure, but without a clearly recognizable preference, the latter (**b**) enabled a significant predominance of the surface area covered by one domain orientation compared to the other one (From [17])

(Sect. 5.5.2). Only due to the different crystallographic direction of the slight misorientation, the identical VPE process led to clearly distinguishable results. For any arbitrary off-cut orientation different from the {011} directions such as about 30° off of [011] on the substrate of type 'S1' shown in Fig. 5.20a, the Si-dimers of the $(2 \times 1)/(1 \times 2)$ surface reconstruction of Si(100) can neither run parallel nor perpendicular to the general orientation of the step edges. Hence, the preparation resulted in a complex surface structure: the AFM image (Fig. 5.20a) shows a configuration of single-layer steps separating triangular shaped terraces arranged in two perpendicular directions. Regarding the surface area covered by either of the two domains, the image does not show any obvious preference.

In contrast, a slight off-cut in one of the {011} directions produces a consistence of the dimer orientations with general direction of the step edges as depicted in Fig. 5.20b for a substrate of type 'S2' which was tilted by about 0.1° almost exactly in [011]: the AFM image (Fig. 5.20b) shows very straight step edges running from top to bottom which differed by two atomic layers in height.

However, many small triangular structures of single-layer height appear to be attached to the straight edges, but they cover only a small fraction of the total surface area. The terraces on the upper side of the straight step edges often appear to reach the next one indicating local double-layer step formation. Accordingly, we employed Si(100) substrates of type 'S2' for the preparation of preferentially double-stepped surfaces with a very low fraction of the residual surface reconstruction domain (Fig. 5.20b) prior to GaP heteroepitaxy.

5.5.2 Growth Conditions for Straight Propagation of APBs

The growth of GaP on Si(100) substrates was undertaken in a two step process, where first of all a homoepitaxial Si buffer was deposited using vapor phase epitaxy (VPE) and the precursor silane and afterwards the III–V semiconductor GaP was grown on this buffer. Detailed growth conditions for the Si buffer layer and its annealing as well as for the GaP nucleation layer subsequently grown on this buffer layer have been given in [8, 22, 36]. In the following, we will briefly summarize the growth conditions resulting in an allocation of the APBs on {011} planes. The metal–organic vapor phase epitaxy (MOVPE) of GaP took place in the Aixtron AIX 200-GFR reactor system using Pd-purified H_2 carrier gas at a reactor pressure of 50 mbar. The metal–organic precursors tertiarybutyl phosphine (TBP) and triethyl gallium (TEGa) were used for the deposition of the layer, also due to the low-temperature decomposition characteristics of these metal organic compounds. Flow-rate modulated epitaxy (FME), where the Ga and P precursor are transported to the reactor intermittently, was applied at a temperature of 450 °C in order to obtain a defect-free nucleation layer as well as a configuration of APBs which reproduces the monoatomic steps at the Si-surface by propagating on {011} lattice planes. The temperature has been calibrated to the Al/Si eutectic formation at 577 °C.

5.5.3 Preparation of the P-rich Surface Reconstruction

Prior to the in situ optical surface characterization, our samples were subject to uncontrolled surface oxidation due to their transport between the remote MOVPE reactors through air. Accordingly, we applied a process for the re-preparation of their surfaces including oxide removal and in situ RAS controlled formation of the well-known P-rich surface reconstruction [6] using purified hydrogen as carrier gas at a base pressure of 100 mbar and the precursor TBP at a partial pressure of about 0.21 mbar. The supply of a surplus of P during the process was essential to the replace the losses due to the formation and evaporation of oxides and to avoid disproportionate P evaporation leading to a Ga-rich termination of the surface [11, 21]. In detail, the reproducible preparation of the P-rich surface reconstruction on the samples required annealing at 650 °C for 5 min to achieve a complete deoxidation and a subsequent cool-down to 300 °C. The TBP supply was switched off at that temperature, but the samples were still covered with excess P which accumulated at the surfaces during cooling. As described in detail in Ref. [6], this situation is a good starting point for the reproducible preparation of the P-rich surface reconstruction, since the excess P-desorption can be monitored easily by in situ RAS. The signal intensity successively recovered during careful heating to 375 °C and stabilized after about 5 min annealing, when all excess P had disappeared from the P-rich reconstructed surface. Finally, the samples were cooled to ambient temperature to exclude influences of the measurement temperature on the quantitative evaluation the RAS data.

5.5.4 In Situ APD Quantification of the GaP/Si(100) Surface

Subsequent to identical Si(100) surface preparation as shown in Fig. 5.20, we deposited thin pseudomorphic GaP films on both substrate types employing special growth conditions [36] for straight propagation of the APBs perpendicular to the surface (Sect. 5.5.2). For a quantitative analysis of the domain structure on the surface [3], both samples (in the following shortly referred to by 'S1' and 'S2' according to the Si(100) substrate type used for preparation) were transferred to a remote MOVPE reactor equipped with in situ RAS. Here, the P-rich surface reconstruction was re-prepared (Sect. 5.5.3) for the analysis. Since the accurate APD quantification versus a principally single-domain GaP(100) surface [6] homoepitaxially prepared in the same reactor required a correction for interference of the reflection at the GaP/Si(100) heterointerface (Sect. 5.3.5), we first recorded the respective relative reflectance spectra shown in Fig. 5.21.

Ensured by the identical surface preparation, we expected exactly the same reflectance for all surfaces, but only the GaP/Si(100) samples featured the buried heterointerface inducing an internal reflection as a source of Fabry-Pérot type interference during measurement. Accordingly, the relative reflectance spectra calculated versus the GaP(100) reference (Fig. 5.21) show significant oscillations below the first direct transition of GaP at about 3.7 eV, and a ratio of approximately unity above, since most of the light was absorbed before reaching the GaP/Si(100) heterointerface. A dielectric model fitted to the optical response of our samples enabled a precise layer thickness determination resulting as expected in very similar values for both GaP/Si(100) samples (Fig. 5.21) which were prepared by identical processes on different types of Si(100) substrates. Beyond that, we required the relative reflectance spectra for the empirical correction [6] of the Fabry-Pérot type interference, which affected the normalization of the in situ RA spectra shown in Figs. 5.22 and 5.23.

In principle, the RA spectrum measured at the sample 'S1' grown on the rather two-domain Si(100) substrate (Fig. 5.22, thin black spectrum) showed the general features attributed to a P-rich reconstructed GaP(100) surface [11, 21] (Fig. 5.22, thick grey spectrum) including the low-energy peak at about 2.4 eV and the high-

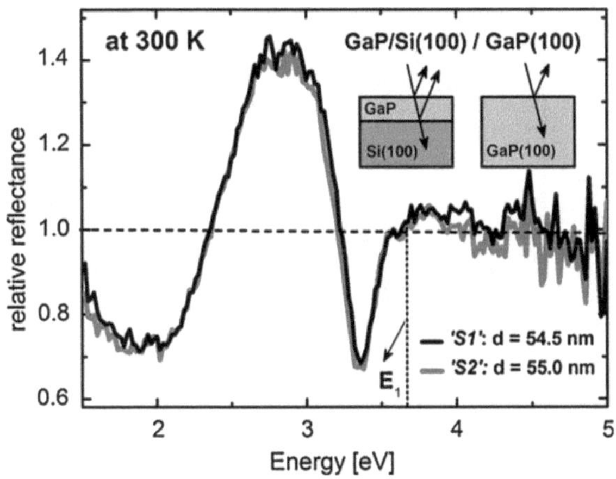

Fig. 5.21 Relative reflectance of heteroepitaxial GaP films grown by the identical process involving conditions for straight APB propagation on different Si(100) substrate of the types 'S1' (*thin black line*) and 'S2' (*thick grey line*). The calculation versus an identically prepared homoepitaxial GaP(100) surface visualizes interferences with the interfacial reflection and allows precise layer thickness determinations by dielectric simulation of the optical response (From [17])

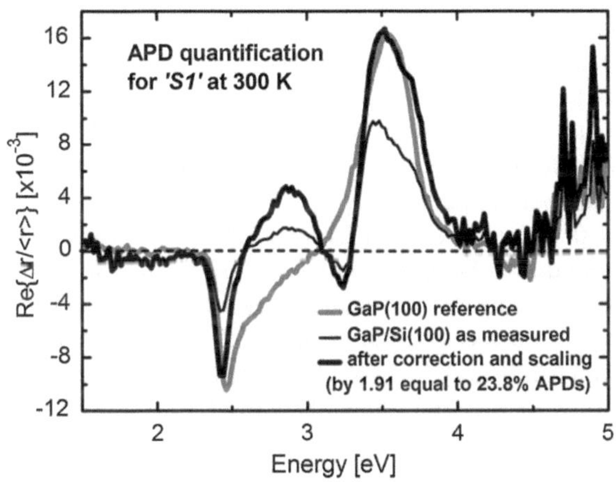

Fig. 5.22 Quantitative in situ RAS analysis of the anti-phase disorder in the GaP/Si(100) film grown on the distinctly two-domain substrate 'S1': the measured data (*thin black curve*) required the correction of interferences via the relative reflectance prior to linear scaling of the signal (*thick black curve*) to the intensity of the homoepitaxial reference signal (*thick grey curve*) eventually resulted in the determination of APD concentration of the sample of about 23.8 % (From [17])

energy peak at about 3.6 eV. However, the observed RAS intensity remained clearly below the values associated to the GaP(100) reference due to the oppositely signed RAS contributions from the APDs [6]. For an accurate quantification of the APD

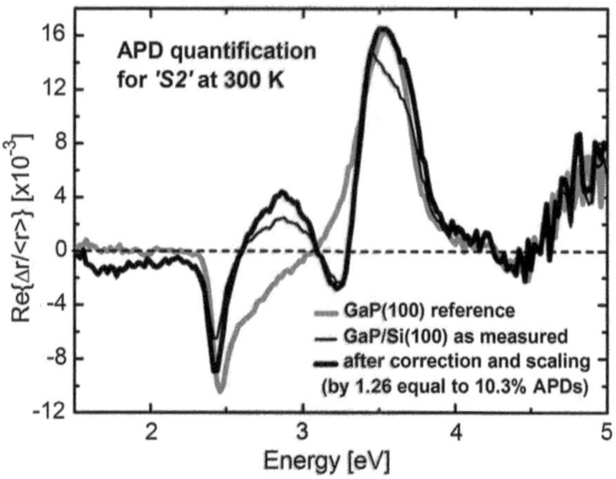

Fig. 5.23 Quantitative in situ RAS analysis of the anti-phase disorder in the GaP/Si(100) film grown on the predominantly single-domain substrate 'S2': the measured data (*thin black curve*) required the correction of interferences via the relative reflectance before linear scaling of the signal (*thick black curve*) to the intensity of the homoepitaxial reference signal (*thick grey curve*) eventually resulted in the determination of APD concentration of the sample of about 10.3 % (From [17])

concentration on the surface, we first convoluted the RA spectrum of sample 'S1' (Fig. 5.22, thin black line) with the respective relative reflectance (Fig. 5.21, black line) according to Eq. 5.2. Linear scaling of the result (Fig. 5.22, thick black line) to fit the intensity of the reference (Fig. 5.22, thick grey line) led to a required factor of $m = 1.91$, which according to Eq. 5.1 indicated about 23.8 % APDs at the surface. In general, the compared spectra (Fig. 5.22) were in good agreement with each other, except for the region between 2.6 and 3.3 eV, where we expect significant signal contributions due to the optical anisotropy of the GaP/Si(100) heterointerface.

In direct analogy to the domain analysis of sample 'S1' (Fig. 5.22), Fig. 5.23 shows the RA spectrum measured at sample 'S2' (thin black line), the GaP(100) reference spectrum (thick grey line) and the fully corrected and scaled result of our evaluation according to Sect. 5.3 (thick black line). In contrast to the other sample, 'S2' was grown on an almost single-domain Si(100) substrate surface (Fig. 5.20b) and therefore should contain less APDs. Actually, the RAS spectrum of 'S2' (Fig. 5.23) was significantly more intense than that of 'S1' (Fig. 5.22) and almost reached the GaP(100) reference without amplification. The full quantitative RAS evaluation of sample 'S2' resulted in a factor of $m = 1.26$, which indicated an APD concentration of about 10.3 % at the surface.

Beyond the trivial result that an improved Si(100) surface structure prior to heteroepitaxy led to a reduction of the anti-phase disorder in subsequently deposited GaP layers, our investigation was designed to enable indirect conclusions about the substrate surface. Due to the applied GaP growth conditions (Sect. 5.5.2), we expected straight propagation of the APBs initiated at single-layer steps of the Si(100) substrates perpendicular to the final surface of our GaP/Si(100) samples. Hence, the APDs should directly trace the domain structure of the Si(100) surface prior to heteroepitaxy without any changes in size or shape. In particular, the in situ RAS quantification of the domain area on the final GaP/Si(100) surface was simultaneously also indirectly valid for surface reconstruction domains of the initial Si(100) surface. In spite of the good qualitative agreement with the direct microscopic analysis in Fig. 5.20, AFM appeared insufficient as quantitative reference since the information was not only hard to acquire, but probably also subject to uncontrolled lateral inhomogeneities of the domain concentration. Instead, detailed structural analysis of the APBs in our heteroepitaxial GaP films by specific dark-field transmission electron microscopy techniques provided further evidence for their straight propagation perpendicular to the surface.

5.5.5 Ex Situ Verification by Dark-Field TEM Analysis

For an in depth TEM investigation of the defect structure of our GaP/Si(100) films [17], {011} cross sectional as well as (100) plan view specimens were prepared from the samples after RAS characterization. Figures 5.24 and 5.25 show TEM micrographs observed under specific dark-field (DF) imaging conditions, which are sensitive to the anti-phase disorder [4] of the samples 'S1' and 'S2', respectively.

In the plan-view TEM DF ($g = (020)$) image of sample 'S1' shown in Fig. 5.24a, the positions of the APBs can be seen from their dark contrast, since diffraction conditions for the plan view case were chosen in a way, that the main phase and the anti-phase have the same bright contrast and only the APBs are visible. All of them appear as thin and sharp line in the on-axis images which confirmed that they run straight through the entire GaP layer—otherwise the dark contrast would be blurred. In Fig. 5.24a sample 'S1' is shown, i.e. a GaP film grown on a Si(100) surface, which was prepared by the identical process and on the same substrate type as the one shown in Fig. 5.20a. It is evident, that the APBs in the GaP film exactly trace the single-layer steps of the underlying substrate; consequently, the APDs in the film actually reproduce the Si(100) surface reconstruction domains prior to heteroepitaxy. In addition, cross-sectional TEM DF ($g = (002)$) images of sample 'S1' are also shown in Fig. 5.24: the specimens were prepared parallel to the [011] (Fig. 5.24b) and to the [011] (Fig. 5.24c) direction, respectively. The diffraction conditions [4] for the cross section dark field micrographs were chosen in a way that two different GaP phases show up with different contrast (dark/bright), while Si has no contrast conditions.

Since the APBs did not necessarily run parallel to the viewing direction, different domains may superimpose over the specimen thickness during DF TEM observation. However, both images (Fig. 5.24b, c) show vertical structures confirming APBs which run straight through the entire GaP layer.

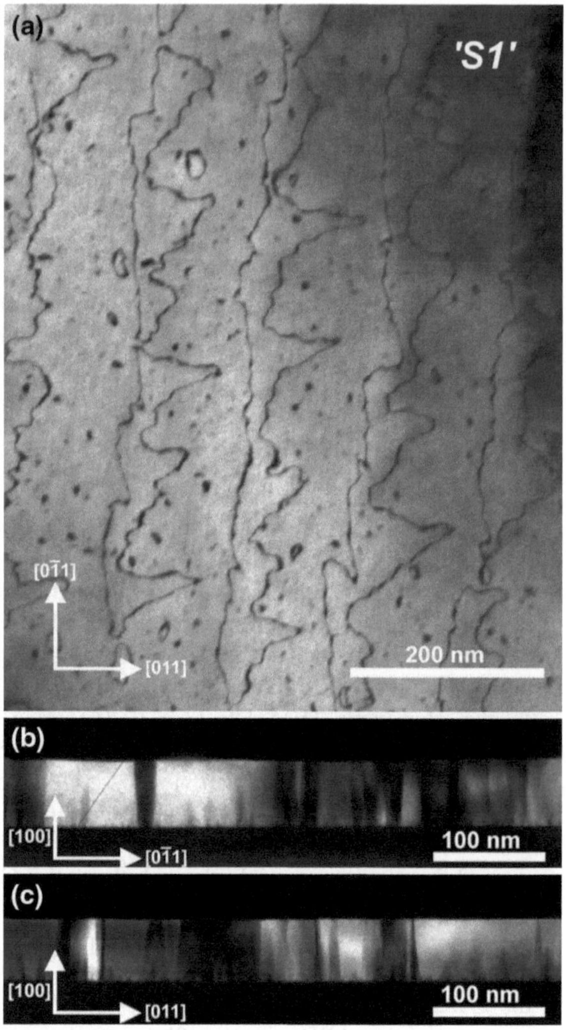

Fig. 5.24 Structural characterization of the sample 'S1' by DF TEM: the imaging conditions for the plan-view (**a**) were chosen to visualize the APBs as *dark lines*, while the two different GaP phases show up with *dark/bright* contrast in the $[0\bar{1}1]$ (**b**) and $[0\bar{1}1]$ (**c**) cross sections, respectively (From [17])

Figure 5.25 shows DF TEM images of the sample 'S2', where—in direct analogy to 'S1'—a GaP layer was grown in FME mode on Si(100). In this case, we used the same Si(100) substrate type and surface preparation as shown in Fig. 5.20b. Again, the anti-phase domain structure visible in the plan-view TEM DF ($g = (020)$) image (Fig. 5.25a) reproduces the atomic surface structure of the Si(100) substrate prior to heteroepitaxy. As expected, the APBs appear as thin and sharp lines. Also the cross-sectional TEM DF ($g = (002)$) image of a sample 'S2' (Fig. 5.25b) confirmed straight APBs running perpendicular to the surface through the entire GaP layer. The specimen for Fig. 5.25b was prepared in the $[0\bar{1}1]$ direction in parallel to the step edges of the Si(100) substrate surface. Hence, the viewing direction of the DF micrograph is perpendicular to the step edges.

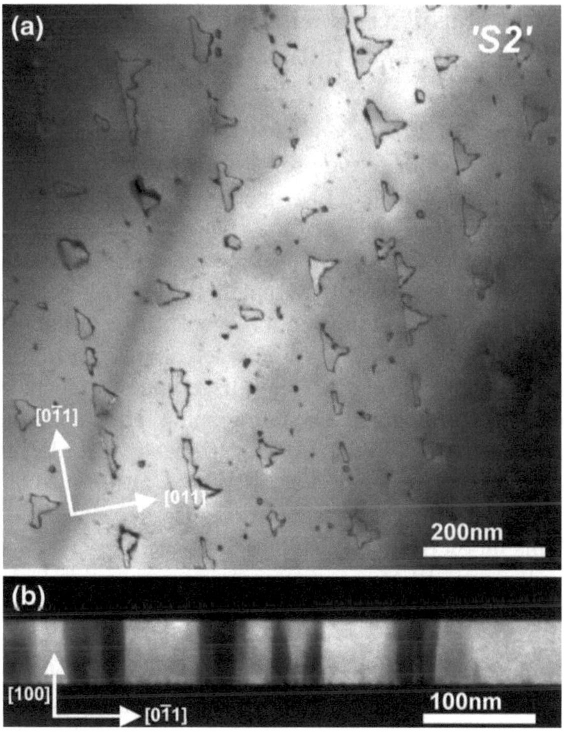

Fig. 5.25 Structural characterization of the sample 'S1' by DF TEM: the imaging conditions for the plan-view (**a**) were chosen to visualize the APBs as *dark lines*, while the two different GaP phases show up with dark/bright contrast in the [0$\bar{1}$1] cross section (**b**) (From [17])

According to the applied diffraction conditions, the two different GaP phases show up with dark/bright contrast and reveal vertical structures confirming straight APB propagation.

Both the cross-sectional as well as in the plan-view DF TEM images shown in Figs. 5.24 and 5.25 contain fundamental verifications for straight APB propagation, which was expected due to the chosen GaP growth conditions (Sect. 5.5.2) and required for the extended interpretation of the in situ APD quantification results (Sect. 5.3.5) as a measure for the concentration of Si(100) surface reconstruction domains prior to heteroepitaxy. The qualitative comparison of the atomic surface structure of VPE-prepared Si(100) substrates as observed by AFM (Fig. 5.20a, b) with the APD structure of GaP films grown on the same substrate types as observed by TEM (Figs. 5.24a, 5.25a) yields very good agreement of the involved microscopic structures. The observation of very small APDs distributed over the terraces by TEM confirmed the absence of APB self-annihilation due to their straight propagation. However, their microscopic origin on the Si(100) surface was not resolved by our ex situ AFM measurements and requires dedicated investigation in future experiments.

5.6 Lateral Resolution of Anti-phase Disorder on III–V/Si(100) Surfaces

Among the established APD characterization techniques, only dark-field TEM provides a domain-type-sensitive resolution of APDs. However, due to the constraints of typical TEM specimen sizes and destructive sample preparation, detailed investigations of the spatial distribution of anti-phase disorder on mesoscopic scales with TEM and a correlation between the microscopic distribution of APDs as resolved by TEM and their overall concentration as measured by RAS are hardly possible. In this section, we describe our application of low energy electron microscopy (LEEM) for a characterization of anti-phase disorder in III–V films grown on Si(100) [15]. The utilized contrast mechanism in LEEM is based on the mutually orthogonal crystallographic orientations of the reconstructed surface domains, which reflect the anti-phase disorder in the GaP film. With this destruction-free method areas of several μm in diameter are easily imaged. Systematic characterization of APD disorder and representative inspection of III–V films on Si(100) have become possible by our new approach.

For our LEEM measurements we grew a thin GaP film of 40 nm thickness on a slightly misoriented Si(100), without suppressing anti-phase disorder in the film. In situ RAS control of the preparation in a MOVPE reactor according to Ref. [6] ensured the formation of the well-defined P-rich surface reconstruction. Directly after deposition, the sample was rapidly transferred to ultra-high vacuum (UHV), loaded into an actively pumped transfer chamber [37] and transported to a remote LEEM where it was introduced via a dedicated load lock. Since the pressure in the sample environment never exceeded the low 10^{-9} mbar range, this effort ensured the protection of the prepared P-rich surface reconstruction against contamination. Eventually, the LEED pattern observed with the microscope [Fig. 5.26 (inset)] after transfer confirmed the expected two-domain surface reconstruction.

The microscope is operated with a parallel electron beam for simultaneous illumination of the whole field of view. The use of electrons with low primary energy in LEEM (0–30 eV) ensures high surface sensitivity [38]. An aperture in the diffraction image plane allows the selection of specific diffraction spots of the observed LEED pattern for the generation of magnified surfaces images. In the present study we used a reconstruction specific contrast mechanism for surface imaging of the heteroepitaxial GaP/Si(100) film: dark-field imaging employing higher-order diffraction spots for the discrimination of the APDs on the surface [39]. For surface imaging of the APD structure we selected the half-order diffraction spots of the respective domain orientation from the LEED pattern (inset of Fig. 5.26) to achieve dark-field contrast between the mutually perpendicularly oriented reconstruction domains. With dark-field imaging we achieved a domain contrast ratio $C = \left|I_{[011]} - I_{[01\bar{1}]}\right| / \left(I_{[011]} + I_{[01\bar{1}]}\right)$ of up to $C = 0.65$. $I_{[011]}$ and $I_{[01\bar{1}]}$ denote LEEM intensities from surface areas with the respective domain orientation and with the terminating P-dimers directed in $[011]$ and $[01\bar{1}]$ direction,

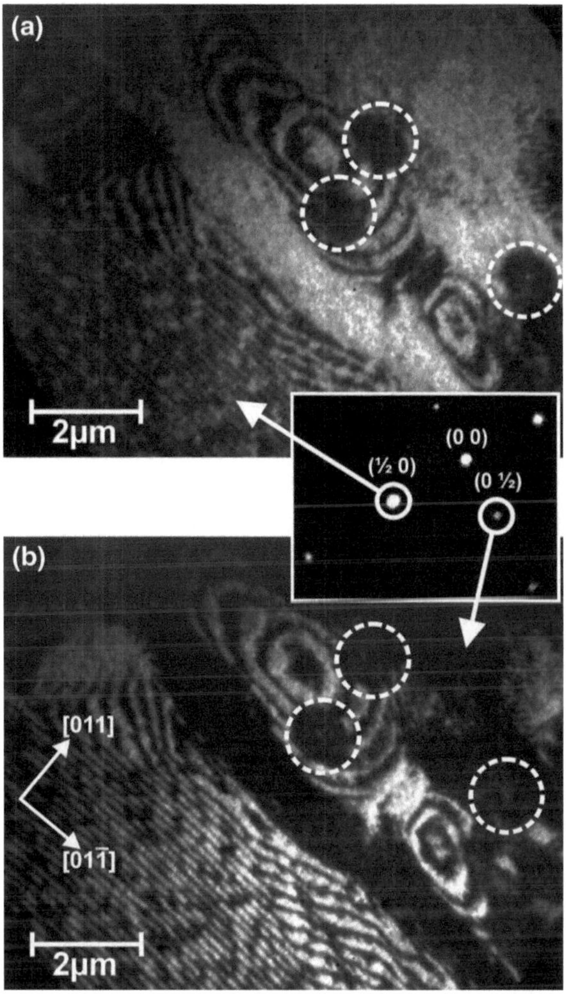

Fig. 5.26 Phase-selective LEEM images ($E_p = 10.2$ eV) of anti-phase disorder in a heteroepitaxial GaP/Si(100) film: *dark-field* imaging with the characteristic (½ 0) (**a**) and (0 ½) (**b**) diffraction reflexes (see inset for LEED pattern at $E_p = 9.7$ eV) results in high contrast of the respective domain orientations. *Encircled spots* can not be attributed to neither of the phases (From [15])

respectively. We expect that this contrast ratio still can be increased to a value of about 1 by suppression of the secondary electron background in our LEEM images.

Figure 5.26 demonstrates the applicability of LEEM for a laterally resolved investigation of APDs in heteroepitaxial III–V layers on Si(100). Changing the imaging conditions by selection of the respective half-order diffraction reflex we obtained nearly complementary images highlighting the surface areas with the P-dimer oriented parallel (Fig. 5.26a) or perpendicular (Fig. 5.26b) to the [011] direction. As expected, the superposition of the bright areas of the two complementary dark-field images covers nearly the entire field of view. Only a few smaller regions of elliptical shape encircled in Fig. 5.26 remain dark independent of the dark-field imaging condition. In the vicinity of these defects a remarkable variation of the lateral APD arrangement in the GaP film occurred in comparison with the majority of the surface far from these spots.

Most regions of the surface appear very similar to the lower left half of the dark-field image pair in Fig. 5.26: the parallel alignment of alternating domains reflects the terrace structure of the substrate. The average domain periodicity of the GaP/Si(100) surface of about 70 nm agrees with the average terrace width of a Si(100) substrate with 0.1° misorientation. The upper right half of the images in Fig. 5.26 shows defects (encircled dark spots) which obviously affect the surface structure in their vicinities: in contrast to the striped terrace pattern far away from the defects, APDs in the proximity of the defects have larger lateral extensions and are arranged in target-like patterns around the defects.

As the alternating sequence of terraces with (2×1) and (1×2) surface reconstructions of the Si(100) substrate initiates the growth of anti-phase disorder in epitaxial GaP films [2], we suspect that the defects responsible for the defect patterns in our LEEM images of the GaP surface have been present on the initial Si(100) surface prior to GaP deposition. Although the dark-field contrasts applied for APD characterization in LEEM and TEM are based on completely different mechanisms, both techniques provide domain–type-sensitive imaging of GaP/Si(100) anti-phase disorder with high contrast. After completing the LEEM measurements, TEM images from the same sample were obtained. Figure 5.27 displays a comparison between LEEM and TEM dark-field images with APD. Both images show very similar structures which are typical for surface areas distant from the defect structures, but at different length scales. The achieved TEM resolution exceeds that of LEEM by almost two orders of magnitude. Consequently, TEM resolves details of the anti-phase distribution much more clearly. However, TEM requires an elaborate and destructive specimen preparation which limits the available field of view drastically. In contrast, LEEM enables a non-destructive sample evaluation by which, in principle, the whole surface area can be accessed. Regarding the contrast mechanism for APD visualization, in TEM one uses phase-selective dynamical scattering within the GaP film [4], while the interpretation of LEEM images relies on the straight-forward identification of APDs via the crystallographic orientation of the local surface reconstruction [39]. Hence, LEEM is applicable for APD investigations at any stage of the deposition beginning from the first monolayers. Compared to scanning probe microscopy, LEEM is a fast imaging method suitable for non-invasive inspection of III–V/Si(100) surfaces and for in situ observation of post-deposition processes such as annealing procedures. In future studies, the suppression of the secondary electron background by an energy filter should improve both phase contrast and image quality significantly. Future experiments will particularly address quantitative analysis of III–V/Si(100) anti-phase disorder on different length scales by LEEM, comparison to in situ RAS results, and a more detailed investigation of defects and their relation to the growth conditions during sample preparation.

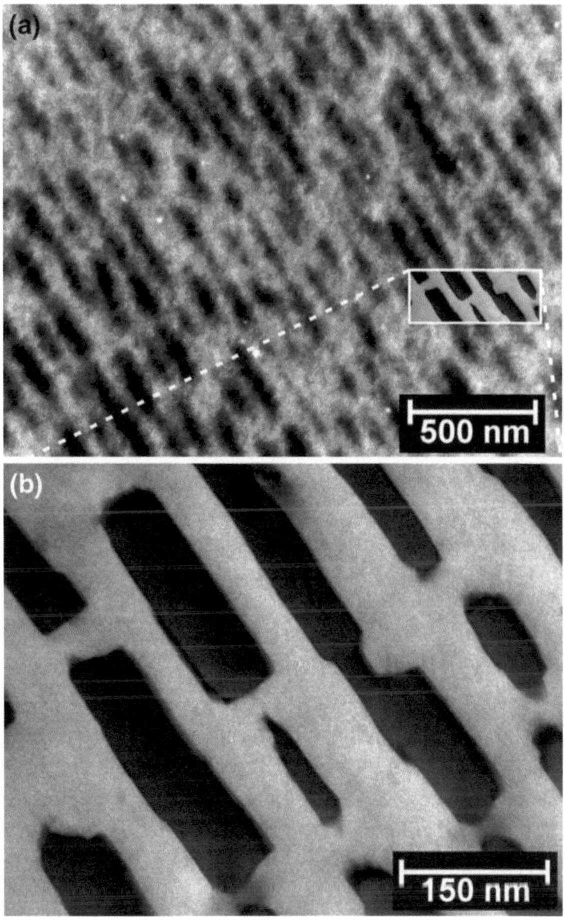

Fig. 5.27 Comparison of a typical LEEM image (**a**) with a plan-view TEM image (**b**) of the same GaP/Si(100) sample, both in a respective *dark-field* mode for the observation of anti-phase disorder. The inset in (**a**) shows a section of the properly scaled TEM image (**b**) superimposed on an arbitrarily selected area of the LEEM image for comparison of the imaged structures (From [15])

5.7 Evaluation of the GaP/Si(100) Heterointerface

The RAS signatures of thin, pseudomorphic GaP films heteroepitaxially grown on Si(100) differ significantly from those observed at identically P-rich prepared GaP(100) surfaces:

1. The presence of anti-phase disorder on the heteroepitaxial GaP/Si(100) surface is expressed by the presence of minority surface reconstruction domains with mutually perpendicular dimer orientation. A linear reduction of the RAS intensity is expected, which can be exploited for in situ quantification of the APD content at the surface.
2. The normalization of GaP/Si(100) RA spectra is subject to interference with an additional reflection induced by the buried heterointerface. Hence, the shape and intensity of the RAS signatures are altered by layer thickness dependent oscillations, but can largely be restored by an empirical correction function.

3. The GaP/Si(100) heterointerface itself potentially defines a supplementary source of optical anisotropy, which could explain further modifications visible in the respective RAS data.

Figure 5.28 shows the comparison of the RA spectra of three GaP/Si(100) samples with different film thicknesses and a homoepitaxial GaP reference spectrum. The preparation of all samples was aimed at ensuring comparability of the RAS data: using the same Si(100) substrate type, absolutely analogous MOVPE processes and the identical scheme for the reliable preparation of the P-rich surface reconstruction and ensuring measurement in a single sequence also including the identically prepared GaP(100) reference surface. The whole effort was dedicated to the investigation of the GaP/Si(100) interface anisotropy by advanced data analysis. In the following, we present an empirical evaluation of the shown RA spectra leading to an estimation of the interface contributions contained in the data on the one hand, and calculations of surface and interface dielectric anisotropies (SDA and IDA, respectively) based on the analysis of our RAS data employing a three-layer model developed by Yasuda [40], which was extended for appropriate consideration of anti-phase disorder.

5.7.1 Empirical Evaluation

The empirical correction of the impact of optical interference on the normalization of RAS signals requires the determination of the relative reflectance spectra of the respective samples in comparison with an adequate reference surface. Figure 5.29 shows the relative reflectance spectra $R_{GaP/Si(100)}/R_{GaP(100)}$ of all three GaP/Si(100) samples in reference to the data of the homoepitaxial GaP(100) surface as well as fitted curves according to optical simulations.

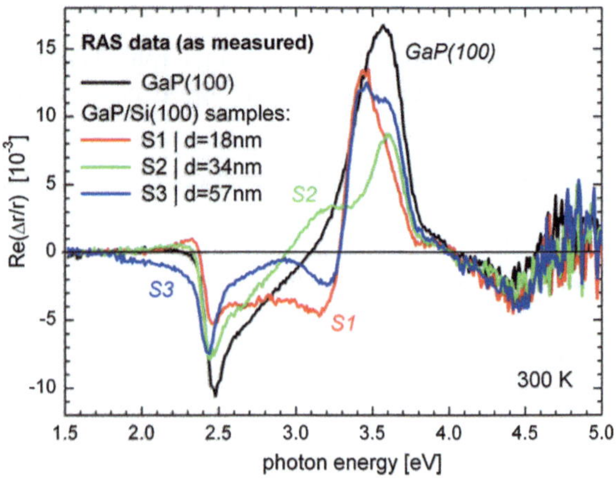

Fig. 5.28 In situ RA spectra of three GaP/Si(100) samples with different GaP film thickness and a homoepitaxial GaP(100) reference. The MOVPE process ensured the formation of a P-rich reconstruction of all surfaces

We used the dielectric data of Ref. [34] and the epitaxial GaP film thickness d_{film} as a parameter to fit the simulation with the experimental data. The resulting layer thicknesses for the samples S1, S2, and S3 are d_1 = 17.7 nm, d_2 = 33.6 nm, and d_3 = 56.7 nm, respectively. The good agreement between experiment and optical simulation confirmed the empirical approach to apply the relative reflectance spectra as correction function for the RAS data.

As an example for the empirical quantification of APDs on III–V/Si(100) surfaces, Fig. 5.30 shows the evaluation of the RA spectrum of sample S3. After application of its relative reflectance spectrum as correction function for interferences in the normalization, the resulting spectrum was scaled to fit the GaP(100) reference spectrum. According to Eq. 5.1 the required amplification factor m determines the APD concentration on the surface. By evaluation of all samples, we determined empirical values of (19.7 ± 3.3) %, (22.5 ± 1.4) % and (19.3 ± 1.2) % for the sample S1, S2, and S3, respectively. The indicated deviation ranges reflect scaling uncertainties associated with imperfect agreement of the shape of the respective sample spectra with the GaP(100) reference. Systematic deviations are clearly visible in Fig. 5.30 and appear to oscillate around the reference spectrum. If these contribution actually could be attributed to GaP/Si(100) interface anisotropy, the oscillating behavior would reflect alternating interference conditions between the optical anisotropy contributions of surface and interface.

Hence, the difference of the fully corrected sample spectra and the GaP(100) reference spectrum has to be considered as a complex derivative of the optically anisotropic interface contribution which is convoluted with experimental errors and modified by the empirical evaluation. Figure 5.31 depicts the spectral difference, which was attributed to our samples by empirical evaluation, and the GaP(100) reference spectrum for comparison of the magnitude. Since the signals are subject to the above-mentioned qualifications and ambiguous signing due to interference, the obtained information remains questionable. However, all spectral differences begin at zero signal for lower energies. The sharp features at about

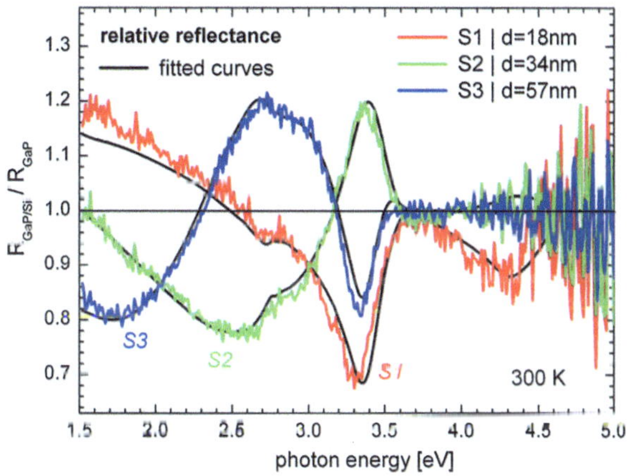

Fig. 5.29 Experimental relative reflectance spectra derived from the data set shown in Fig. 5.28 (*colored lines*) and thickness dependent fit by an optical model (*black lines*)

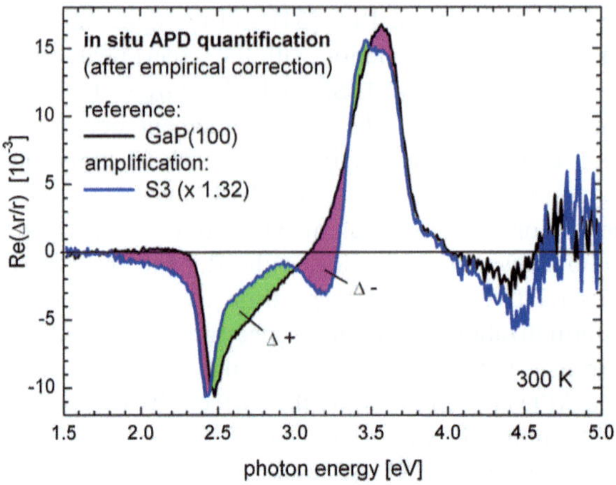

Fig. 5.30 Empirical in situ RAS APD quantification applied to sample S3. After correction of the measured RA spectrum using the corresponding relative reflectance spectrum (Fig. 5.29), empirical amplification of the intensity (*blue*) to match the intensity of the GaP(100) reference (*black*) estimates the APD concentration of the sample. Residual deviations of the sample spectrum (*violet and green*) potentially indicate optical anisotropy of the GaP/Si(100) heterointerface

2.4 eV might be due a strain-induced shift of the first peak of the surface signal. Starting at about 2.7 eV, the absolute value of all signal appears to rise up to a sharp change of the sign at about 3.3 eV. Beyond the onset of absorption at the critical point energy of GaP at about 3.7 eV less light reaches the GaP/Si(100) heterointerface and the signal is largely determined by noise. Since the described behavior could either be an indication about the actual anisotropy of the GaP/Si(100) or just pure coincidence, we started optical calculations to derive the proper interface signal from the identical set of data.

Fig. 5.31 Deviation of the GaP/Si(100) RAS spectra shown in Fig. 5.28 from the GaP(100) reference spectrum (*black*) after full empirical correction

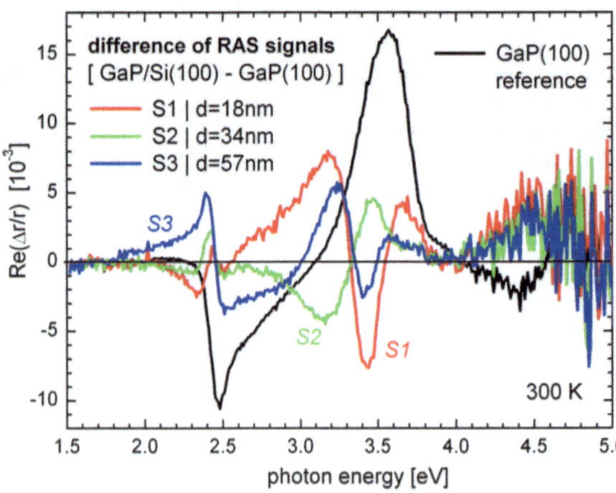

5.7.2 Optical Calculations

In general, RA spectra $\Delta r/r$ measured at a thin film structure of thickness d_{film} contain signal contributions from different sources of optical anisotropy, including those from the surface $\Delta r_{sf}/r_{sf}$, the interface $\Delta r_{if}/r_{if}$ and the bulk Δn of the film. The dielectric model of Yasuda [40] suggests to express the anisotropies as small deviations from the (complex) three-layer reflection coefficient r. First calculations (not shown here) confirmed a negligible contribution of the bulk anisotropy Δn of our GaP films on Si(100). Hence, we express the RAS signal according to Ref. [41] in terms of dielectric anisotropy attributed to the surface and to the interface:

$$\begin{aligned} \frac{\Delta r}{r} &= A(d_{film}) \cdot (d\Delta\varepsilon)_{sf} + B(d_{film}) \cdot (d\Delta\varepsilon)_{if} \\ A(d_{film}) &= \frac{4\pi}{\iota\lambda}\frac{n_0}{\varepsilon_{film} - \varepsilon_0} \cdot \frac{r_{sf}}{r}\frac{\partial r}{\partial r_{sf}} \\ B(d_{film}) &= \frac{4\pi}{\iota\lambda}\frac{n_{film}}{\varepsilon_s - \varepsilon_{film}} \cdot \frac{r_{if}}{r}\frac{\partial r}{\partial r_{if}} \end{aligned} \tag{5.9}$$

Both the surface dielectric anisotropy (SDA, $(d\Delta\varepsilon)_{sf}$), and the interface dielectric anisotropy (IDA, $(d\Delta\varepsilon)_{if}$) are unknown variables. Due to the thickness of our heteroepitaxial GaP films on Si(100), we can assume that the anisotropies located at the surface and interface are well-separated and independent of the layer thickness d_{film}. Hence, a set of two of our samples differs only in the value of d_{film} and defines an equation system base on Eq. 5.9 for each one, which can be solved to derive both anisotropy contributions $(d\Delta\varepsilon)_{sf}$ and $(d\Delta\varepsilon)_{if}$ without fitting or further assumptions about their physical origin.

5.7.2.1 GaP/Si(100) Surface Anisotropy

However, in contrast to the materials studied in Ref. [40], we have to consider the effect of anti-phase disorder induced by III–V/Si(100) heterointerfaces. As discussed above for the empirical model, APDs on the GaP/Si(100) surface reduce the intensity of the RAS signal due to mutually perpendicular dimer orientations of the P-rich reconstruction and lateral integration on a cm scale associated with the measurement [6]. Therefore, the approach was extended by accounting for the APD concentration by a linear scaling factor in direct analogy to Eq. 5.1 [18]. In contrast to the empirical model used so far, we do not apply the amplification to the total measured spectra, but more correctly only to the SDA by replacing

$$A(d_{film}) \;\rightarrow\; A_{sf}(d_{film}, m) = m^{-1} A(d_{film}) \tag{5.10}$$

As a first approach, we use the empirically obtained APD-concentration as additional input parameters for the calculation of SDA and IDA according to

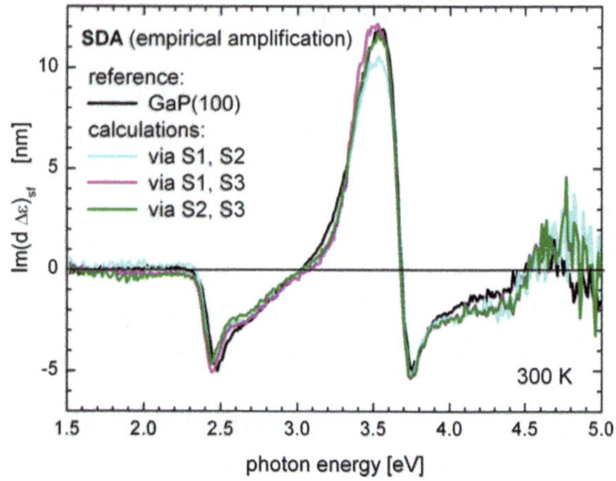

Fig. 5.32 Calculated surface dielectric anisotropy (*SDA*) of GaP/Si(100) derived by calculation of Eq. 5.9 for sets of two samples and input of empirical APD amplification factors in comparison to the reference SDA of GaP(100)

Eq. 5.9 employing all three possible combinations of input RAS data measured at the samples S1, S2, and S3.

Figure 5.32 shows the imaginary parts of the resulting SDA spectra, which should correlate with the measured RAS data, since $\mathrm{Re}(\Delta r/r) \propto \mathrm{Im}(d\Delta\varepsilon)$ for $d \ll \lambda$ [42]. As expected according to the identical preparation of the well-order P-rich surface reconstruction, the shape of the SDA is very similar for all three data set combinations and agrees very well with the SDA $(d\Delta\varepsilon)_{GaP}$ of the measured homoepitaxial GaP(100) reference.

5.7.2.2 Calculation of the APD Content of the GaP/Si(100) Surfaces

The confirmation of the consistency between the SDAs $(d\Delta\varepsilon)_{sf}$ calculated from the samples with $(d\Delta\varepsilon)_{GaP}$ of the GaP(100) reference enabled us to carry out further modifications of our calculation approach. For a direct derivation of the APD concentrations, we kept the APD concentrations of the samples variable, which is expressed by $(d\Delta\varepsilon)_{sf} = (d\Delta\varepsilon)_{sf}(m_k, m_l)$ using sample specific amplification factors m. Giving the equation system of twice Eq. (5.9) for different sets of samples $\{k, l\}$, we determined both factor $\{m_k, m_l\}$ simultaneously by numerical minimization of the difference $(d\Delta\varepsilon)_{GaP} - (d\Delta\varepsilon)_{sf}(m_k, m_l)$ by a least square fit. We expect more precision by this approach, since amplification of the SDA only correctly reflects the physical situation of mutually perpendicular surface reconstruction domains.

In addition, the concept qualifies for implementation as automated algorithm for deriving the APD concentrations at the GaP/Si(100) surfaces without any empirical input. Averaging the output APD concentrations for each sample, we obtained $(22.5 + -1.3)$ %, $(25.8 + -0.9)$ % and $(13.0 + -0.1)$ % for S1, S2, and S3, respectively.

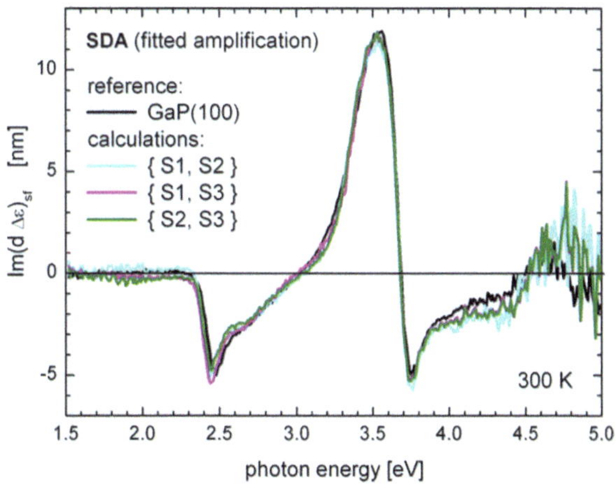

Fig. 5.33 Calculated surface dielectric anisotropy (*SDA*) of GaP/Si(100) derived by calculation of Eq. 5.9 for sets of two samples and fit of the output SDA to that of GaP(100) using the APD concentration of the respective samples as fitting parameters

Figure 5.33 shows the resulting SDAs, which were simultaneously obtained by the calculation of the APD concentrations, for all three possible data combinations in comparison to the GaP(100) reference. Despite a slight red shift of the position of the first peak relative to the homoepitaxial reference, very good agreement of the SDAs in Fig. 5.33 was obtained by the fitting procedure. We suppose the red shift to be caused by the accumulation of strain in our pseudomorphic GaP films, which potentially shifts the energy gap of the anisotropic surface state transition that is associated with this peak [13]. We currently plan to support further investigations of this effect by measurement of an extensive data set containing samples with several different GaP/Si(100) film thicknesses. In general, we obtained very good agreement of the SDAs in Fig. 5.33 and the small variance of the APD concentrations when calculated by different data sets. Hence, the results confirm correct determination of GaP/Si(100) SDAs $(d\Delta\varepsilon)_{sf}$ by our modified approach considering anti-phase disorder.

5.7.2.3 GaP/Si(100) Interface Anisotropy

Corresponding to the SDA spectra shown in Fig. 5.33, the calculation simultaneously resulted in IDA spectra, which are shown in Fig. 5.34. In all cases, the IDAs attributed to each data set of data consisting of the RA spectra of two samples, respectively, feature a small, increasing contribution between about 2.3 and 3.1 eV. Beyond the critical point energy, the noise of all IDA spectra increases significantly due to the onset of adsorption, which affected samples with higher film thickness most. While the IDA line shapes for all data sets are in good agreement, the amplitude of the IDA differs.

To obtain more detailed results for the GaP/Si(100) IDA, we applied further modifications to our calculation approach, which correspond to findings gained so far. Up to this point, we used the GaP(100) for correct scaling only, but in the following, we replace the SDA of the samples $(d\Delta\varepsilon)_{sf}$ by the SDA of the

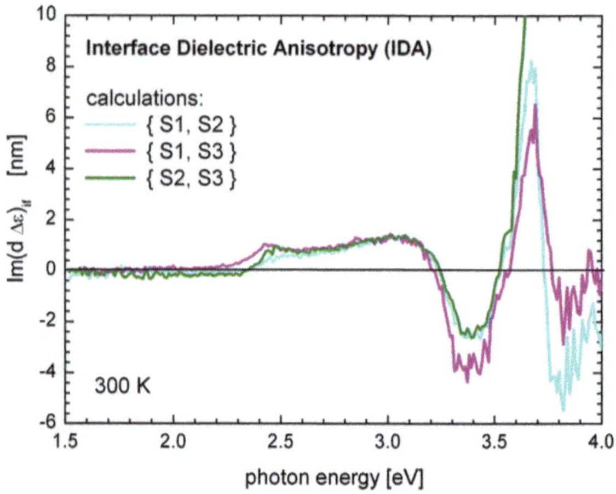

Fig. 5.34 Calculated interface dielectric anisotropy (*IDA*) of GaP/Si(100) simultaneously derived by calculation of the SDA data shown in Fig. 5.33. Due to excessive noise in the IDAs caused by absorption above the critical point energy of GaP at about 3.7 eV, the data is only shown up to 4.0 eV and one spectrum (*green*) had to be cut even earlier

homoepitaxial GaP(100) reference $(d\Delta\varepsilon)_{GaP}$, which enables us to solve Eq. 5.9 directly with the RAS data of a single GaP/Si(100) sample and allows us to derive the corresponding IDA individually:

$$(d\Delta\varepsilon)_{if} = \frac{1}{B}\left(\frac{\Delta r_{GaP/Si}}{r_{GaP/Si}} - m^{-1}A \cdot (d\Delta\varepsilon)_{GaP}\right) \tag{5.11}$$

Since we do not calculate the corresponding SDA, this approach does not allow deriving APD concentrations. Instead, we use the values obtained via the least square fit leading to the results shown in Figs. 5.33 and 5.34 as input parameters.

The obtained IDA results are depicted in Fig. 5.35. Again, the small oscillations around the energetic position of the minimum RAS peak at about 2.4 eV most probably are caused by a small red shift of the SDA of the GaP/Si(100) surface relative to the GaP(100) reference, which is not considered in our calculations so far. The effect appears more pronounced in Fig. 5.35 compared to Fig. 5.34 due to the different input data (sample and reference vs. two samples with slightly different film thickness). Apart from that feature, the IDA attributed to samples S1 and S3 agree fairly well, while the IDA of sample S2 appears to feature significantly lower intensity. Experimental errors and imperfections of the data processing such as the baseline correction are issues, which may cause significant distortion of the IDAs the complex evaluation. However, if anti-phase disorder also affects the optical anisotropy of the interface, linear scaling of the IDA in analogy to the SDA could occur. Based on this assumption, the data shown in Fig. 5.35 could directly indicate the quality of the GaP/Si(100) interface preparation of the different samples, and, in particular the APD concentration at the interface. Actually, the in situ RA spectra of the monohydride terminated Si(100) surfaces prior to GaP heteroepitaxy (not shown here) appear to support this hypothesis, which requires extensive validation by future experiments.

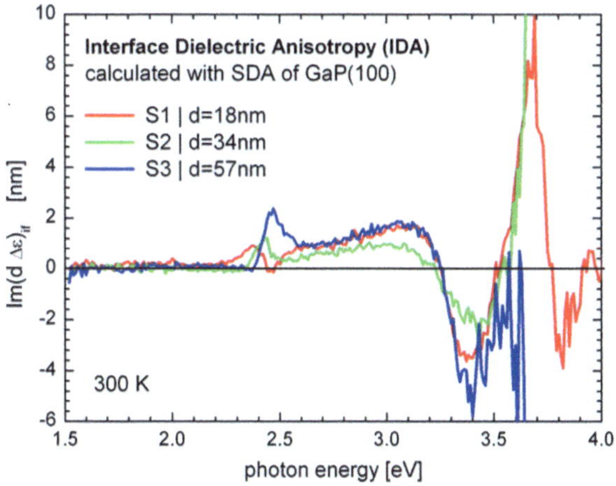

Fig. 5.35 Calculated interface dielectric anisotropy (*IDA*) of the individual GaP/Si(100) samples shown in Fig. 5.28. The IDAs were derived by solving Eq. 5.9 using the GaP(100) reference SDA, individual RAS spectra of the respective samples and the derived amplification factors as input parameters. Due to excessive noise in the IDAs caused by absorption above the critical point energy of GaP at about 3.7 eV, the data is only shown up to 4.0 eV and two spectra (*green and blue*) had to be cut even earlier

References

1. S.F. Fang, K. Adomi, S. Iyer, H. Morkoc, H. Zabel, C. Choi, N. Otsuka, J. Appl. Phys. **68**, R31–R58 (1990)
2. H. Kroemer, J. Cryst. Growth **81**, 193–204 (1987)
3. H. Döscher, T. Hannappel, B. Kunert, A. Beyer, K. Volz, W. Stolz, Appl. Phys. Lett. **93**, 172110 (2008)
4. I. Nemeth, B. Kunert, W. Stolz, K. Volz, J. Cryst. Growth **310**, 4763–4767 (2008)
5. M. Kawabe, T. Ueda, Jpn. J. Appl. Phys. Part 2: Lett. **26**, L944–L946 (1987)
6. H. Döscher, T. Hannappel, J. Appl. Phys. **107**, 123523 (2010)
7. P.N. Uppal, H. Kroemer, J. Appl. Phys. **58**, 2195–2203 (1985)
8. I. Nemeth, B. Kunert, W. Stolz, K. Volz, J. Cryst. Growth **310**, 1595–1601 (2008)
9. J.M. Zhou, H. Chen, F.H. Li, S. Liu, X.B. Mei, Y. Haung, Vacuum **43**, 1055–1057 (1992)
10. J.P. Gowers, Appl. Phys. Mater. Sci. Process. **34**, 231–236 (1984)
11. L. Töben, T. Hannappel, K. Moller, H.J. Crawack, C. Pettenkofer, F. Willig, Surf. Sci. **494**, L755–L760 (2001)
12. N. Esser, W.G. Schmidt, J. Bernholc, A.M. Frisch, P. Vogt, M. Zorn, M. Pristovsek, W. Richter, F. Bechstedt, T. Hannappel, S. Visbeck, J. Vac. Sci. Technol. B **17**, 1691–1696 (1999)
13. P.H. Hahn, W.G. Schmidt, F. Bechstedt, O. Pulci, R. Del Sole, Phys. Rev. B. **68**, 033311 (2003)
14. R.A. Wolkow, Phys. Rev. Lett. **68**, 2636–2639 (1992)
15. H. Döscher, B. Borkenhagen, G. Lilienkamp, W. Daum, T. Hannappel, Surf. Sci. Lett. **605**, L38 (2011)
16. H. Döscher, B. Kunert, A. Beyer, O. Supplie, K. Volz, W. Stolz, T. Hannappel, J. Vac. Sci. Technol. B. **28**, C5H1–C5H6 (2010)

17. H. Döscher, O. Supplie, S. Brückner, T. Hannappel, A. Beyer, J. Ohlmann, K. Volz, J. Cryst. Growth **315**, 16–21 (2011)
18. O. Supplie, T. Hannappel, M. Pristovsek, H. Döscher, Phys. Rev. B. **86**, 035308 (2012)
19. T. Yasuda, D.E. Aspnes, D.R. Lee, C.H. Bjorkman, G. Lucovsky, J. Vac. Sci. Technol. A-Vac. Surf. Films **12**, 1152–1157 (1994)
20. P. Weightman, D.S. Martin, R.J. Cole, T. Farrell, Rep. Prog. Phys. **68**, 1251–1341 (2005)
21. T. Hannappel, L. Toben, K. Moller, F. Willig, J. Electron. Mater. **30**, 1425–1428 (2001)
22. B. Kunert, I. Nemeth, S. Reinhard, K. Volz, W. Stolz, Thin Solid Films **517**, 140–143 (2008)
23. O. Hunderi, J.T. Zettler, K. Haberland, Thin Solid Films **472**, 261–269 (2005)
24. U. Rossow, D.E. Aspnes, Phys. Status Solidi A-Appl. Res. **177**, 157–163 (2000)
25. S. Visbeck, T. Hannappel, M. Zorn, J.T. Zettler, F. Willig, Phys. Rev. B. **63** (2001)
26. D.E. Aspnes, J.P. Harbison, A.A. Studna, L.T. Florez, J. Vac. Sci. Technol. A-Vac. Surf. Films **6**, 1327–1332 (1988)
27. C.T. Foxon, B.A. Joyce, R.F.C. Farrow, R.M. Griffith, J. Phys. D-Appl. Phys. **7**, 2422–2435 (1974)
28. L.N. Glurdzhi, A.P. Izergin, Z.N. Kopylova, A.D. Remenyuk, Sov. Phys. Semiconductors-Ussr. **7**, 305–306 (1973)
29. S.E. Acosta-Ortiz, A. Lastras-Martínez, Phys. Rev. B. **40**, 1426 (1989)
30. O. Acher, S.M. Koch, F. Omnes, M. Defour, M. Razeghi, B. Drevillon, J. Appl. Phys. **68**, 3564–3577 (1990)
31. S. Adachi, J. Appl. Phys. **66**, 6030–6040 (1989)
32. J.T. Zettler, Prog. Cryst. Growth Charact. Mater. **35**, 27–98 (1997)
33. S. Zollner, M. Garriga, J. Kircher, J. Humlicek, M. Cardona, G. Neuhold, Thin Solid Films **233**, 185–188 (1993)
34. D.E. Aspnes, A.A. Studna, Phys. Rev. B. **27**, 985–1009 (1983)
35. A.R. Laracuente, L.J. Whitman, Surf. Sci. **545**, 70–84 (2003)
36. K. Volz, A. Beyer, W. Witte, J. Ohlmann, I. Németh, B. Kunert, W. Stolz, J. Cryst. Growth (2010) (these proceedings, submitted)
37. T. Hannappel, S. Visbeck, L. Toben, F. Willig, Rev. Sci. Instrum. **75**, 1297–1304 (2004)
38. E. Bauer, Rep. Prog. Phys. **57**, 895–938 (1994)
39. H. Marbach, G. Lilienkamp, H. Wei, S. Gunther, Y. Suchorski, R. Imbihl, Phys. Chem. Chem. Phys. **5**, 2730–2735 (2003)
40. T. Yasuda, Thin Solid Films **313**, 544–551 (1998)
41. D.E. Aspnes, J. Vac. Sci. Technol. B, **3**, 1498–1506 (1985)
42. J.D. McIntyre, D.E. Aspnes, Surf. Sci. *24*, 417-& (1971)

Chapter 6
Conclusion

The formation of anti-phase disorder during MOVPE growth of pseudomorphic GaP films on Si(100) substrates served as a lattice matched model system for the crucial III–V/Si(100) interface. A variety of surface-sensitive methods was applied to establish suitable Si(100) substrate preparation and subsequent GaP growth free of anti-phase domains (APDs), by analyzing the substrate surface, the interface and the epitaxial film resulting from the heteroepitaxial process. We studied thermal removal of native and protective wet-chemical SiO_2 layers from Si(100) substrates in a MOVPE reactor by means of XPS and required annealing at a surface temperature of 950 °C for reliable deoxidation in a hydrogen ambient. This critical surface temperature for oxide removal was significantly higher than that in established UHV procedures and could be decreased by lowering the process pressure or by supplying silane, while a slight increase in argon-based processes was observed and the alternative use of nitrogen as process gas induced silicon nitride formation. Using LEED, STM and AFM, we found distinct two-domain $(2 \times 1)/(1 \times 2)$ reconstructions with an equal distribution of smooth S_A type and frayed S_B type monolayer steps on Si(100) samples with 0.1° and 2° off-cut. In contrast, the prevalence of a single domain and a significant fraction of double layer D_B steps were detected on 6° surfaces. We were able to obtain Si(100) surface structures with D_A-like steps by the growth of a homoepitaxial buffer layer even on 0.1° misoriented samples. However, in most places, high resolution STM images revealed intermediate S_B steps only a few dimers long. Dependent on the rate of cooling, we also observed D_A-like steps on the 2° Si(100). Although intermediate S_B steps were visible, in several places true D_A steps seemed to prevail. Since our findings contradict established UHV results for Si(100) surfaces with and without hydrogen coverage, we investigated the interaction of hydrogen from the process ambient with the Si(100) surfaces. ATR mode FTIR spectra showed the characteristic coupled stretch modes of H–Si–Si–H dimers and tip-induced STM desorption experiments confirmed a complete monohydride termination after processing. However, in situ RAS characterization at typical surface preparation temperatures indicated effectively hydrogen free Si(100), since desorption processes probably outbalance hydrogen adsorption.

H. Döscher, *GaP heteroepitaxy on Si(100)*,
Springer Theses, DOI: 10.1007/978-3-319-02880-4_6,

The in situ analysis of heteroepitaxial GaP films on Si(100) requires a detailed understanding of the surface reconstruction mechanisms involved. The atomic structure of P-rich reconstructed GaP(100) typical for MOVPE preparation was characterized by STM. The images resolved rows of alternatingly buckled H-stabilized P-dimers arranged either in phase or out of phase, which refer to (2×2) and $c(4 \times 2)$ domains of the surface reconstruction, respectively. The investigations revealed a dimer flipping mechanism, which could also be found on P-rich prepared InP(100). Flipping events switch the tilt of individual dimers and may locally leed to a transformation between the two possible P-rich surface reconstructions. In general, GaP(100) and InP(100) surfaces behave very similar, in particular during MOVPE preparation using hydrogen as the standard process gas, but we found and characterized an additional surface reconstruction regime for preparation nitrogen preparation of GaP(100), which does not exist for InP(100). To permit quantitative signal analysis, the impact of changes of the sample temperature as well as the reconstruction and the atomic order of the GaP(100) surface on the assigned RAS signatures were considered in separate experiments.

Optical in situ spectroscopy was established as a method for the quantitative evaluation of the APD content in GaP heteroepitaxy. The evaluation of the APD content of a distinctly two-domain GaP/Si(100) sample in reference to another heteroepitaxial GaP film of equal thickness led to promising agreement with ex situ AFM measurements. In contrast, RA spectra of undoubtedly single-domain GaP(100) were not directly applicable as reference surface, since interference with additional reflections at the GaP/Si(100) heterointerface alters shape and intensity of the sample spectra. As shown by detailed signal analysis, the major share of the characteristic deviations originates from incorrect normalization of anisotropic surface signals. The determination of the relative reflectance of a sample compared to a P-rich prepared GaP(100) reference surface enables application of corresponding corrections of the RAS data of heteroepitaxy experiments to obtain an accurate APD quantification independent of the GaP film thickness. Extensive benchmarking versus microscopically resolving and statistically evaluated ex situ reference techniques such as AFM and dark-field TEM confirms the reliability and accuracy of the in situ RAS quantification of III–V/Si(100) anti-phase disorder. Using specific film growth conditions for straight propagation of APBs, we simultaneously obtain an indirect in situ quantification of Si(100) surface reconstruction domains, since the APD structure of a GaP/Si(100) surface directly reproduces the distribution of surface structure of the Si(100) substrate prior to heteroepitaxy as evidenced by combined AFM and dark-field TEM ex situ measurements. However, bridging the gap between the local information of such microscopically resolving APD characterization techniques on the one hand and the macroscopic information of laterally integrating surface characterization techniques such as in situ RAS and LEED on the other hand usually requires extensive statistics. LEEM dark-field imaging provides an alternative for direct evaluation of III–V/Si(100) anti-phase disorder on a mesoscopic length scale with a non-destructive surface sensitive microscopy technique. Subsequent to their full empirical correction with respect to interference in the normalization and anti-phase disorder, systematic deviations between the RA spectra GaP/Si(100) and GaP(100) samples were still observed, in particular in the spectral range around 3.2 eV. The characteristic differences were attributed to the optical anisotropy of the GaP/Si(100) heterointerface. We conducted complex model calculations, for which the approach of Ref. [1] was extended for the consideration of anti-phase disorder, to separate the contribution of the surface and interface dielectric anisotropies (SDA and IDA, respectively) to the RA spectra of our samples.

Current experiments address several various aspects of the present work. The unconventional surface structure of VPE prepared Si(100) featuring D_A double-layer steps is of highest fundamental and technological interest. The reproduction of the STM results with variation of the process parameters during preparation might reveal details of the formation mechanism. For technological application in the field of III–V heteroepitaxy it is important to investigate the concentration of the residual minority domain and the reliability of the preparation. With regard to the quantitative in situ evaluation of anti-phase disorder, transfer to a different III–V on IV will be an important step, conceivable materials combinations are AlP/Si, GaAs/Ge and GaInP/Ge, of which the latter is probably of highest interest with regard to device applications, in particular for multi-junction photovoltaics. Also of fundamental interest is the actual, microscopic structure of the III–V/Si(100) interface formed during nucleation. Perhaps, the optical model calculations for accurate separation of surface and interface contribution to the optical anisotropy observed by RAS will result in some indications, in particular when combined with advanced interface sensitive non-linear optical spectroscopy techniques such as second harmonic generation. However, thorough surface characterization might also contribute significant insight into this issue. Most of the techniques provide information about the dimer orientation of the initial Si(100) surface prior to heteroepitaxy or of the final GaP/Si(100), some even both. In a very simple model, the relation of both dimer orientations signifies the type of polarity at the interface established during nucleation. Last, but not least, the application of high quality GaP/Si(100) films as a quasi-substrate for device structures is of highest interest, but requires either the development of sufficient grading strategies or the utilization of challenging dilute nitride materials lattice matched to Si.

Reference

1. T. Yasuda, D.E. Aspnes, D.R. Lee, C.H. Bjorkman, G. Lucovsky, J. Vac. Sci. Technol. A-Vac. Surf. Films **12**, 1152–1157 (1994)

MIX
Papier aus verantwortungsvollen Quellen
Paper from responsible sources
FSC® C105338
FSC
www.fsc.org

If you have any concerns about our products,
you can contact us on
ProductSafety@springernature.com

In case Publisher is established outside the EU,
the EU authorized representative is:
Springer Nature Customer Service Center GmbH
Europaplatz 3, 69115 Heidelberg, Germany

Printed by Libri Plureos GmbH
in Hamburg, Germany